International Web Conference on Advances in Sustainability of Materials and Environment (IWCASME 2020)

Held online due to COVID-19

IOP Conference Series: Materials Science and Engineering
Volume 983

Tamil Nadu, India
17 - 18 September 2020

Editors:

J. Jerlin Regin
A. Suhasini

ISBN: 978-1-7138-3180-8
ISSN: 1757-8981

Printed from e-media with permission by:

Curran Associates, Inc.
57 Morehouse Lane
Red Hook, NY 12571

Some format issues inherent in the e-media version may also appear in this print version.

Institute of Physics
Dirac House, Temple Back
Bristol BS1 6BE UK

Phone: 44 1 17 929 7481
Fax: 44 1 17 920 0979

techtracking@iop.org

Additional copies of this publication are available from:

Curran Associates, Inc.
57 Morehouse Lane
Red Hook, NY 12571 USA
Phone: 845-758-0400
Fax: 845-758-2633
Email: curran@proceedings.com
Web: www.proceedings.com

TABLE OF CONTENTS

IOP Conf. Series: Materials Science and Engineering **983** (2020) 011001 doi:10.1088/1757-899X/983/1/011001

PREFACE

This proceeding comprises selected papers from the 3rd International Web Conference on Advances in Sustainability of Materials and Environment (IWCASME 2020) conducted on 17th & 18th September 2020. It covers a broad range of technical and review articles within the topics Environmental Science, Civil and Building Science, Energy Efficient Technologies, Mathematical Modeling, Chemistry and Physical Science and Material Science.

The existing pandemic situation made the organizers to conduct the conference through online. The conference was a combined endeavour by the Department of Civil Engineering and the Department of Chemistry of St. Xavier's Catholic College of Engineering, Chunkankadai, Nagercoil, Kanyakumari District, Tamil Nadu, India.

The conference focused on the interaction between theoretical and practical applications in the given areas. Interactive online presentations were arranged through Google Meet for the participants to present their ideas. Four sessions and eight venues were arranged for paper presentation. All the selected papers were peer reviewed by expert reviewers in a single- blind review system as per the review policy given by IOP Conference series: Material Science and Engineering.

On 17th September Invited talks were presented by Dr.Md.Abdul Mannan, Professor, University Malaysia Sarawak, Malaysia on Sewage Sludge- A Renewable and Sustainable Resource in Formulating Lightweight Aggregate for Precast Concrete Industry and Dr. Benny Joseph, Principal, Vimal Jyothi Engineering College, Kannur- Kerala, India on Sustainability and COVID-19. On 18th September Dr.M.Asha Jhonsi, Assistant Professor (Sr.Gr.), Crescent Institute of Science and Technology, Chennai, Tamil Nadu, India delivered a talk on Environmental Waste to Carbon Nanomaterial- An Overview Towards Sustainable Development and Dr.S.Ganapathy Venkatasubramanian, Professor, Anna University, Chennai, Tamil Nadu, India presented a talk on The Role of Legal Regime for the Sustenance of Sustainable Development in India.

The conference committee acknowledges and appreciates the strenuous efforts taken by the presenters, authors, reviewers and the publisher IOP Conference series: Material science and Engineering. It is hoped that the forum enriched the knowledge of research scholars, industrialists, faculty members and students.

IOP Conf. Series: Materials Science and Engineering **983** (2020) 011001 doi:10.1088/1757-899X/983/1/011001

EDITORS:

1. Dr.J.Jerlin Regin (HOD/CIVIL, SXCCE)

2. Dr.A.Suhasini (Assistant Professor/Chemistry, SXCCE)

List of committees:

1. Advisory technical committee

1. Dr.Md.Safiuddin, MACI,FIEB-Toranto

2. Dr.L.Leon Raj, CSIR-NEIST Jorhat-Assam

3. Dr. C.Rajasekaran,NIT- Manglore

4. Dr. S.Nagan, Thiyagarajar College of Engg- Madurai

5. Dr.G.Rexin Thusnavis, PKSC- Nagercoil

6. Dr.P.Maria Pushpam, PKSC- Nagercoil

7. Dr.P.Vincent, Mepco Schlenk Engg.College-Virudhunagar

8. Dr.M.Berlin, NIT-Arunachal Pradesh

9. Dr.A.B.Danie Roy, Thappar University-Punjab

10. Dr.T.P.D.Rajan, NIIST-Trivandrum

11. Dr.K.P.Vinod Kumar, UCEN- Nagercoil

12. Dr.S.VasuDevan,CSIR-Karaikudi

Organizing Committee

1.Dr.J.Maheswaran, (Principal, SXCCE, Organizing Chairman)

IOP Conf. Series: Materials Science and Engineering 983 (2020) 011001 doi:10.1088/1757-899X/983/1/011001

2.Dr.J.Jerlin Regin (HOD/CIVIL,SXCCE, Organizing Secretary)

3.Dr.A.Suhasini (Assistant Professor/Chemistry, SXCCE, Organizing Joint Secretary)

4.Dr.S. Carmel Jawahar (Associate Professor/ CIVIL, SXCCE, Convenor)

5.Dr.T.M.Chenthil Jegan (Coordinator/ HSC,SXCCE, Convenor)

Programme Committee

1. Mrs. A.MariaSheela (Assistant Professor/ Chemistry, SXCCE)
2. Dr.A.M.Alice Margret (Assistant Professor/ Chemistry, SXCCE)
3. Mrs. S.A.Anuja (Assistant Professor/ Chemistry, SXCCE)
4. Mrs.S. Shiela Balanta (Assistant Professor/ CIVIL, SXCCE)
5. Mr.P. Antony Vimal (Assistant Professor/ CIVIL, SXCCE)
6. Mr.M. Galesh (Assistant Professor/ CIVIL, SXCCE)
7. Mr.M. Inigo Valan (Assistant Professor/ CIVIL, SXCCE)
8. Mr.S. Frank Stephen (Assistant Professor/ CIVIL, SXCCE)
9. Mr.E. Lourdu Amalan (Assistant Professor/ CIVIL, SXCCE)
10. Mr.J. Jeya Suren Raj (Assistant Professor/ CIVIL, SXCCE)
11. Mr.F. John Paul(Assistant Professor/ CIVIL, SXCCE)
12. Mr.M. Joseph David Selvan (Assistant Professor/ CIVIL, SXCCE)
13. Mrs.L. Porcia (Assistant Professor/ CIVIL, SXCCE)
14. Mrs.D. Smiline Shiny(Assistant Professor/ CIVIL, SXCCE)

Publication Committee

IOP Conf. Series: Materials Science and Engineering **983** (2020) 011001 doi:10.1088/1757-899X/983/1/011001

1. Dr. I.Jessy Mol, Assistant Professor/Civil, SXCCE

2. Dr.V.Sreeja, HOD/ Chemistry, SXCCE

IOP Conf. Series: Materials Science and Engineering **983** (2020) 011002 doi:10.1088/1757-899X/983/1/011002

Peer review declaration

All papers published in this volume of *IOP Conference Series: Materials Science and Engineering* have been peer reviewed through processes administered by the Editors. Reviews were conducted by expert referees to the professional and scientific standards expected of a proceedings journal published by IOP Publishing.

- **Type of peer review: Single-blind with two level review**
- **Conference submission management system: Research Papers were submitted using conference mail id: iwcasme20@gmail.com**
- **Number of submissions received: 51**
- **Number of submissions sent for review: 35**
- **Number of submissions accepted: 17**
- **Acceptance Rate (Number of Submissions Accepted / Number of Submissions Received X 100): 33.33%**
- **Average number of reviews per paper: 2**
- **Total number of reviewers involved:16**
- **Any additional info on review process :Nil**
- **Contact person for queries: Dr. J. Jerlin Regin,**

Associate Professor and Head,
Editor IWCASME 20,
Department of Civil Engineering,
St.Xavier's Catholic College of Engineering,
Chunkankadai,
Kanyakumari District- 629003,
Tamil Nadu, India.
Email id: jerlin@sxcce.edu.in

IOP Conf. Series: Materials Science and Engineering **983** (2020) 012001 doi:10.1088/1757-899X/983/1/012001

A performance assessment on Compressed Stabilized Earth Block (CSEB) using geo – polymeric binders

Sangeethavani B[1], Muthukannan M[2], Deepak R[3], Arun Pradeep A[4]

[1]**Assistant** Professor, Centre for Rural Technology, The Gandhigram Rural Institute – Deemed to be University.
[2] Professor, Department of Civil Engineering, Kalasalingam Academy of Research and Education, Krishnankoil, India
[3,4]Final Year Students, B.Tech Civil Engineering, Centre for Rural Technology, The Gandhigram Rural Institute – Deemed to be University.

Abstract. The development of the country mainly depends on Rural Development. About 55% of rural peoples still live in earth building. According to the US Green Building Council, the building sector accounts for 39% of the global emission of CO_2. If new technologies were not adopted means the emission could be doubled by the year 2050 according to United Nation Environment Program (UNEP).New technology is required to sustain the environment and earth materials. In India, One of the recently developing technology is Compressed Stabilized Earth Block (CSEB). This CSEB technology replaces the usage of conventional burnt clay bricks. In this Context, The earth materials are not widely used in CSEB because it includes the vital part of industrial wastes. The characteristics of CSEB with Granulated Blast Furnace Slag (GGBFS), Fly Ash of class-F and coir fiber are to be investigated. Nowadays, the controlling, management and disposal of waste materials are a difficult task. The Flyash-F is an industrial waste material that is obtained from a thermal power plant and it contains a low amount of CaO. GGBFS is also an industrial waste obtained from the quenching molten iron slag in steel industries. About 20-25% of dry volume soil is replaced by a combination of Flyash and GGBFS with the addition of a geopolymer solution. Both the waste materials bring the CSE-Block as eco- friendly to the environment and cost- effective. Besides, The CSEB exhibits high strength, low creep, very minimum water absorption and good erosion resistance. In this paper, we are experimentally investigating the compressive strength and water absorption nature of CSEB with varying ratios of binders and Geo-Polymeric solution.

1. Introduction

Mud based technology was an ancient technology and it was started during the ninetieth century. A Compressed Stabilized Earth Block also is known as stabilized Earth block. It was stabilized by a different stabilizing agent such as Fly ash, Ground Granulated Blast Furnace Slag (GGBFS) and Geo- Polymeric solution such as Sodium hydroxide and Sodium silicate (Na2SiO3) as Activating Alkaline Solution (AAS). CSEB technology developed for environmentally friendly and cost- effective construction.

The Flyash class-f pulverized from the bituminous coal. At present, NLC (Neyveli Lignite Co- operation) produces 200 million tons of Flyash per year. GGBFS in-organic waste material from steel industries. The effect of Efflorescence was reduced when GGBFS was using about 50-60% and it protects against chloride and sulfate attack. The Geo-polymeric solution was In this project, The combination of Flyash and GGBS will increase the compressive strength of CSEB and decrease the water absorption of the

Block. The optimum percentage of GGBS (20%) gives the maximum increment in the CBR value. Red soil sample and15% Flyash and 25% GGBFS give compressive strength was 14.83 MPa at 28 days. The water binder ratio was 0.52 (T.V. Preethi May 2014). Experimental investigations about Flyash, GGBS, Silica fume and Kadapa slab powder. 75% flyash+22%GGBS+3% silica fume with 8M gives excellent compressive strength about 10.3MPa at 28 days and water absorption was 6.7%. 75% flyash+22%GGBS+3%kadappa slab powder gives the rate of water absorption was 6.6% (Sunandana Reddy Mar-Apr 2018).

In this experiment an attempt was made to produce a new innovative and cost effective building material to achieve high strength of CSE- Block by using industrial waste material. The optimum dosage of GGBFS and Fly Ash was obtained. The experimental investigations on the effect of altering important variables such as coir fiber, GGBFS and fly ash content on stabilized earth blocks was done. In addition the compressive strength of CSEB with different curing methods was also performed. The result shows that the developed technology was able to produce high strength, innovative and cost effective building material.

2. Materials & Methodology

The methodology includes the theoretical principles and analysis of the body associated with a branch of knowledge.

2.1. *Red soil*

Red soil was procured locally from Dindigul. The red soil is rich in iron content and it is the key factor for enhancing the strength in the block. The engineering properties are determined using the standard tests as per I.S codes.

Table 2.1. Properties of soil

Parameters	Values
Gravel	10%
Sand	15%
Silt	36%
Clay	39%
Liquid limit, LL	15.65%
The plastic limit, PL	7.5%
Plasticity index, PI	8.15%
Optimum Moisture Content	15%
Maximum Dry density (ρ_{max})	2.33g/cm^3
Maximum Bulk density (ρ_{bulk})	2.03g/cm^3
Specific gravity, G	2.23

2.2.Flyash – F

Flyash class-F is obtained by a coal-based thermal power station procured from Tuticorin. Specific Gravity of Flyash was 2.78. The Flyash material sieved to ensure the material free from lumps. The Specific gravity test conducted by as per IS 1727- 1967.

Figure 1. Flyash-F

2.3 Ground Granulated Blast Furnace Slag (GGBFS)

GGBFS contains 30-50% of lime content. So, the combination of Flyash and GGBFS increases the strength of the block. GGBFS was procured from Madurai. Specific Gravity of GGBFS was 3. The test was done as per Indian standards IS 1727-1967.

Figure 2. GGBFS

2.4. Coir Fiber

The coir fiber was procured from the local market at free of cost. The fibers are cut into 3 to 4cm. The coir fiber was treated before the block making. It was soaked in a NaOH solution for 2 hours and it was dried 70°C in the oven for about 5hrs.

Figure 3. Treated coir fiber

2.5. Alkaline solution

The Alkaline solution also is known as the Geo-Polymer solution. The NaOH solution prepared with a concentration of 8M & 10M. The molecular weight of NaOH was 40. The mass of NaOH solids was measured as 255 grams per kg of NaOH solution for 8M concentration. The solution was prepared at normal room temperature. When the solution mixed, both solution starts to react (polymerization) and it liberates a large amount of heat so it is recommended to leave it for about 24 hours thus the alkaline solution gets ready as a binding agent.

Figure 4. Alkaline solution

2.6. Mix proportions

The CSE-Block mix proportion is investigated with two different Molarity 8M & 10M. The binder ratio (NaOH: Na_2SiO_3) was 1:2. Average Weight of one block was 6 Kg. actually; 0.6 kg of the alkaline solution is required for the making of one block. The results are discussed on Table 3.1 & 3.2. The soil percentage is taken as base constant (75%) and molarity of Geo-Polymer solution fixed as 8M in the main mix proportion based on experimental investigation.

Table 2.2. Trial mix proportions

Mix ID	Red soil (%)	Flyash –F (%)	GGBFS (%)	Coir fiber (%)
M1	75(8M)	12	12	1
M2	75(10M)	12	12	1
M3	75(8M)	12.5	12.5	-
M4	80(8M)	10	10	-
M5	85(8M)	7.5	7.5	-

Table 2.3. Main mix proportions

Mix ID	Red soil (%)	Flyash-F (%)	GGBFS (%)
CSEB-1	75	5	20
CSEB-2	75	10	15
CSEB-3	75	15	10
CSEB-4	75	20	5
CM	75	-	25

2.7. Casting & Curing Of CSEB

The stabilized mud block was made by using the HYDRO FORM machine. The blocks were made with a Geo polymer solution at various proportions. The size of a block is 210 x 115 x 115 mm. Initially, the dry mix was done by adding GGBS & Flyash with soil and then the alkaline solution was added and allowed to mix with help of hydraulic rotating drum.

Figure 5. HYDRO FORM machine

The block should be taken gently and they should be kept on a smooth and flat surface at 24hours for the Resting period of block. Curing was carried out with different curing methods such as oven and ambient curing. CSE-Blocks placed on the oven at 60^0C for 24hrs. Water curing doesn't require for Geo-Polymer blocks.

Figure 6. Oven Curing

3. Result and discussions

3.1. Compression test: (IS3945:1992)

Compressive strength is the ability of structure or material to carry the loads on its surface without any crack or deflection. Compressive strength of CSE-Block depends on many factors such as water- binder ratio, the molarity of an alkaline solution, quality of GGBFS & Flyash-F material, quality, etc... The area of the block was 224.2cm^2. The compression test was conducted on the 3rd, 14th, 28th day of curing. The wet and dry compressive Strength of CSE-Block in both oven and ambient curing was tested at the end of 28 days.

Table 3.1. Compression value at 14th day of curing

S.no	Sample ID	Average Compressive Strength (MPa)
1	M1	9.56
2	M2	8.49
3	M3	16.03
4	M4	7
5	M5	8.7

Table 3.2. Compression value at 28th day of curing

Sample ID	Dry Compressive Strength (Mpa)		wet Compressive Strength (MPa)
	Ambient Curing	Oven Curing	
CSEB- 1	23.72	26.7	14.95
CSEB- 2	23.82	24.71	11.37
CSEB- 3	12.09	13.72	8.50
CSEB- 4	13.53	17.12	10.82
CM	24.62	28.0	16.44

IOP Conf. Series: Materials Science and Engineering **983** (2020) 012001 doi:10.1088/1757-899X/983/1/012001

Figure 7. Compression test

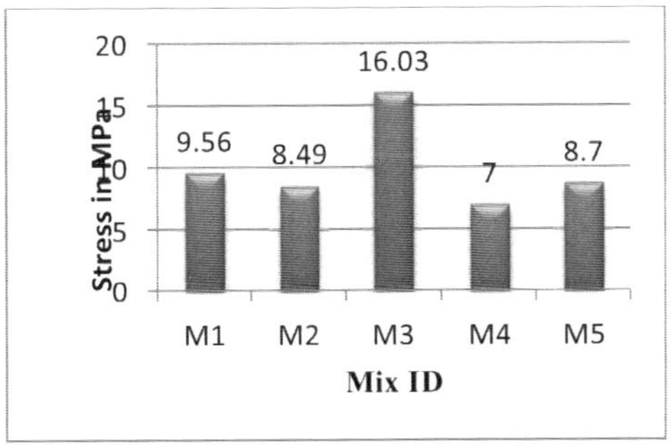

Figure 8. Compressive strength at 14th day of curing

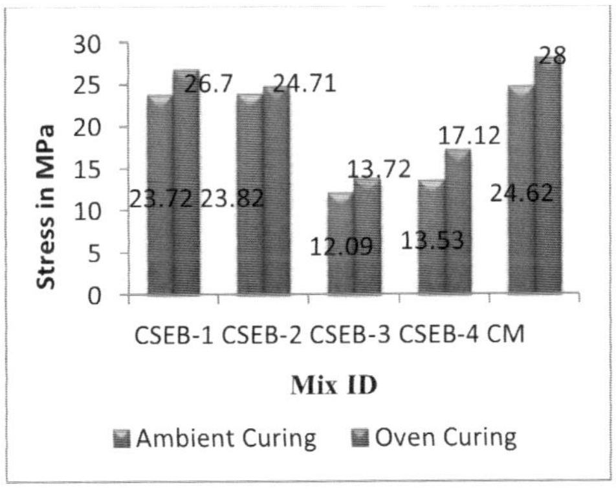

Figure 9. Dry Compressive strength at 28th day of curing

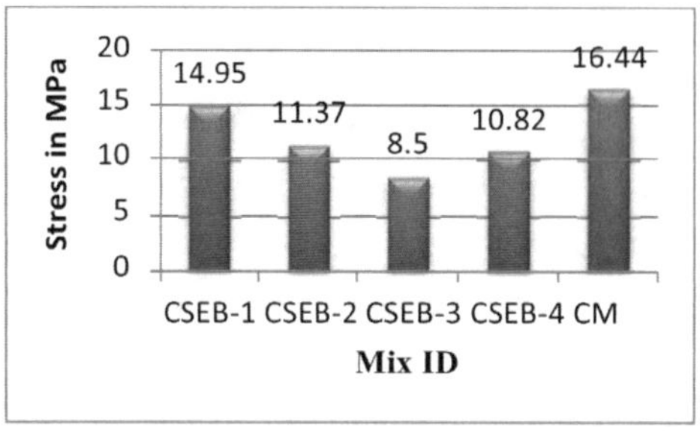

Figure 10. Wet Compressive strength at 28th day of curing

3.2. *Water absorption test: (is 3495:1992(part-II))*

Water absorption of CSE-Block is an important factor in classifying its durability. Generally, First class bricks having a high water absorption of about 20%. Stabilized mud block is less than 13% as per Indian standards.

Figure 11. Water absorption test

Table 3.3. Water Absorption Test

Mix ID	Dry Weight (Kg)	Wet Weight (Kg)	Water Absorption (%)	Permissible limit
CSEB- 1	5.734	5.9	2.81	
CSEB- 2	5.562	5.73	3.02	
CSEB- 3	5.435	5.74	5.31	
CSEB- 4	5.73	6.02	5.06	
CM	5.862	6.1	4.06	
M1	5.56	5.93	6.65	
M2	5.74	6.07	5.75	

M3	5.86	6.09	3.92
M4	5.62	5.87	4.45
M5	6.02	6.3	4.65

Water absorption can be measured by dry weight and wet weight of the specimen. It was expressed in a percentage of the mass of the sample. The minimum water absorption rate was 2.81%. The mix proportion of CSEB-1 gives better results.

3.3.*UPV Test: (Is 13311:1992 Part-1)*

An ultrasonic pulse velocity (UPV) test is an in-situ, non-destructive test to check the quality of concrete and mud blocks. In this test, the strength and quality of mud block are assessed by measuring the velocity of an ultrasonic pulse passing through a mud CSE-blocks. Higher velocities indicate good quality and continuity of the material, while slower velocities may indicate block with many cracks or voids. The test procedure is done according to the standard code specification of IS1331 (part1):1992.

Table 3.4. UPV Test Result

S.no	Mix ID	Receiving point	Pulse Velocity (L/ T) in Km/s
1	CSEB-1	2HZ	4.32
2	CSEB-2	2HZ	3.03
3	CSEB-3	2HZ	3.9
4	CSEB-4	2HZ	3.78
5	CM	2HZ	4.32

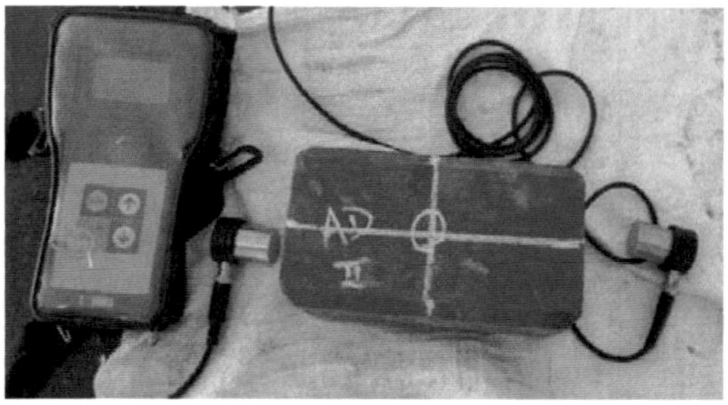

Figure 12. UPV Test

3.4. Scanning Electron Microscopy (SEM) analysis

It provides high-resolution imaging useful for evaluating various materials for surface fractures, flaws, contaminants. It gives results about the physical and chemical properties of the sample. It helps to identify the chemical composition of the sample.

Table 3.5. Chemical composition of Flyash-F

Element	Normalized Concentration[wt.%]
Ca	11.39
O	46.24
Si	23.52
Al	7.71
Fe	9.54
C	1.09
Mg	0.51

Figure 13. Microscopic view Flyash

Table .3.6. Chemical composition of GGBFS

Element	Normalized Concentration[wt.%]
Ca	11.18
O	39.85
Si	34.41
Al	7.92
Mg	2.54
C	0.21
S	0.64

Figure 14. Microscopic view GGBFS

4. Conclusion

From the study of CSEB with Geo- Polymeric binders. The conclusion is based on the experimental study on compressive strength and water absorption of CSE-blocks. We concluded the following points:

- The Optimum Percentage of soil is taken as 75% of the results from the trial mix proportion. Because increment in the percentage of soil leads to achieving unsatisfied results.
- The strength of block decreases when the molarity of the solution increases.
- The strength of the block increases when the flyash content increases in the range of 5-10% gradually. But when the fly ash content increases more than 10%, the block strength reduces up to 10MPa.
- The CSE-Block made of proportion was 75% of red soil+20% of GGBFS +5% of Flyash. It gives better geometrical properties and yield better compressive strength on different curing methods. The stress was 26.7 MPa at 28 days (oven curing) and 23.72 MPa at 28 days (Ambient curing). The rate of water absorption was 2.87%.
- The Proportion of 75% of red soil+5% of GGBFS +20% of Flyash. The compressive strength was 17.12 MPa at 28 days (oven curing) and 13.53 MPa at 28 days (Ambient curing). The rate of water absorption was 5.06%.
- If Flyash content increases the rate of absorption of the block also increases.
- The ultrasonic pulse value range between 3 to 4.7 as an average. It indicates good internal

quality and flawless composition of CSE-B block.

Reference

[1] Oormila TR and Preethi TV 2014 Effect of the Stabilization Using Flyash and GGBS in Soil Characteristics *International Journal of Engineering Trends and Technology (IJETT)* **11** 284-289.

[2] Palanisamya B, Suresh Kumar P 2018 Effect of molarity in geopolymer earth brick reinforced with fibrous coir wastes using sandy soil and quarry dust as fine aggregate *Case Studies in Construction Materials* **8** 347–358.

[3] Sunandana Reddy, Lakshmi Usha, Anwar Basha, Abhinay, Guru Bhaskar and Kusuma,"An Experimental Investigation on Hybrid Geopolymer bricks by using Fly Ash, GGBS, Silica Fume and Kadapa Slab powder".

[4] Vinu Prakash, Amal Raj, Aravind S, Basil Mathew and Sumith VR "STUDIES ON STABILIZED MUD BLOCK AS A CONSTRUCTION MATERIAL" International Journal of Innovative Research in Advanced Engineering (IJIRAE) | ISSN: 2349-2763

[5] IS3495 (PART-1&2):1992, "Methods of tests of burnt clay building bricks".

[6] IS 1727:1967, "Methods of test for pozzolanic materials".

[7] IS 2720:1983, PART –3, "Methods of test for Soil".

[8] IS 2720:1985, PART –5, "Determination of liquid and plastic limit".

A review on the emergence of single-chamber microbial fuel cell on wastewater treatment

R Jinisha[1]*, J Jerlin Regin[2], J Maheswaran[2]

[1] Department of Civil Engineering, University College of Engineering, Nagercoil, TamilNadu, 629004, India
[2] Department of Civil Engineering, St. Xavier's Catholic College of Engineering, Chunkankadai, TamilNadu, 629003, India

Corresponding e-mail: r.jinisha8@gmail.com

Abstract. The principle of generating electrical energy using biomass forms the basis of Microbial Fuel Cells (MFCs). This new technology not only produces electricity but also cleans up the waste. Based on the number of chamber units, MFCs are mainly divided into single chamber (SCMFC) and double chamber (DCMFC). This paper reviews particularly about SCMFC. The fundamental components of SCMFC like anode and cathode and the various microbes used in the fuel cell are explained in this review. This paper details about the materials used for the synthesis of anode and cathode. Also, the applicability of SCMFC for the purification of synthetic and real wastewater is discussed.

1. Introduction

Energy exists in different forms such as chemical energy [1], light energy [2], heat energy [3], mechanical energy [4], gravitational energy [5], electrical energy [6], sound energy [7], nuclear or atomic energy [8] and so on. Of all these energies, electrical energy plays a vital role in our routine life [9,10]. With the increasing population of the world, the demand for energy is also increasing which leads to an energy crisis [11–13]. For the past years, the energy was mostly obtained from fossil fuels [14–16]. Now there is depletion of these fuels since they are non-renewable. So, energy from renewable energy sources is preferable [17]. One such renewable energy source is fuel cell [18]. It is an open thermodynamic equipment in which electrical energy is produced from chemical energy [19,20]. Among the existing types of fuel cells, biofuel cells are the most popular [21]. They are minuscule fuel cells that employ biological organisms as catalysts for the production of electrical energy [22,23]. Biofuel cells are broadly classified into enzymatic biofuel cells, in which enzymes of living organisms are used and the entire microbial cell of organisms is used for oxidation of fuel in microbial fuel cells. Among the mentioned types of biofuel cells, Microbial Fuel Cells (MFCs) are largely utilized because of two major advantages that involve production of clean electrical energy and treatment of wastewater. It has been used widely in the treatment of sludge [24], saline wastewater [25], dairy industry wastewater [26], Purified terephthalic acid [27] wastewater from the petrochemical industry, urine wastewater [28], slaughterhouse wastewater [29], livestock wastewater [30], dyestuff wastewater [31] and so on. The other use of MFC is as a biosensor it is employed in the online wastewater quality monitoring [32]. MFC technologies are able to convert chemical energy directly to electricity through a biological pathway [33]. It can generate clean electricity directly and also there is no necessity for purification, segregation and energy product aleration [34]. It is a bioreactor in which chemical energy that is available in the chemical bonds gets converted to electricity via the reactions of anaerobic microorganisms which act as catalysts [35]. One such example for a microorganism which can generate electricity and also accomplish treatment of wastewater is bacteria [36,37]. The vital components of an MFC are an anode compartment, a cathode compartment, an external circuit of wire which connects anode and cathode, a Proton Exchange Membrane (PEM) and microorganisms [38]. Anode and cathode electrodes are placed in anode and cathode compartments respectively. In an MFC, the community of microbes present in the anode compartment catalysts the oxidation process of the organics that is provided in the anode chamber. When this oxidation process takes place, protons, electrons and also CO_2 are generated [39]. The resultant electrons are moved to the cathode chamber through an external electric circuit, while protons are moved to the cathode chamber through a membrane called proton exchange membrane (PEM) [40]. This PEM allows the diffusion of the protons which are produced in the anode compartment to the cathode compartment. In this cathode area, water is formed by the reaction between protons, electrons and oxygen. Oxygen will not allow electricity generation. Therefore, in the anode chamber, the anaerobic condition is maintained compulsory. But, the cathode chamber is exposed to oxygen to enhance water formation [41]. The reactions that take place in both the chambers are shown in equations 1 and 2 [42].

At anode electrode: $CH_3COO^- + 2H_2O \longrightarrow 2CO_2 + 7H^+ + 8e^-$ (1)

At cathode electrode: $O_2 + 4H^+ + 4e^- \longrightarrow 2H_2O$ (2)

MFCs have two categories of classification depending on the number of chambers as single chamber microbial fuel cells (SCMFC) and two chamber/dual chamber microbial fuel cells (DCMFC) [43]. The main difference is that the former MFC consists of only one anode compartment with no definite cathode compartment [33,44] while the latter has both anode and cathode compartments [45–47]. Many review papers have been published regarding MFC and its developments till date. But this paper reviews only the SCMFC which is simple and cheap [48].

2. Fundamental Features of SCMFC

The illustration of simple SCMFC is shown in Figure 1. The principle of the design of this reactor is similar to a hydrogen fuel cell. In hydrogen fuel cells, there is direct contact between the cathode and PEM. The cathode takes oxygen from the atmosphere. In single compartment microbial fuel cells (SCMFC), there is only one anode chamber in which wastewater to be treated is placed along with the microbial catalysts. It is maintained in anaerobic condition for the survival of anaerobic microorganisms present there. External wire conducts the electrical ions to the cathode that is generated in the anode. In SCMFC, air is directly exposed to the cathode, thus taking oxygen from the atmosphere [49]. Sometimes, to separate the two electrodes, Gas Diffusion Layer (GDL) or PEM is attached to the cathode. The basic features of SCMFC are the two electrodes - anode and cathode, micro-organisms and membrane [50] which are explained below.

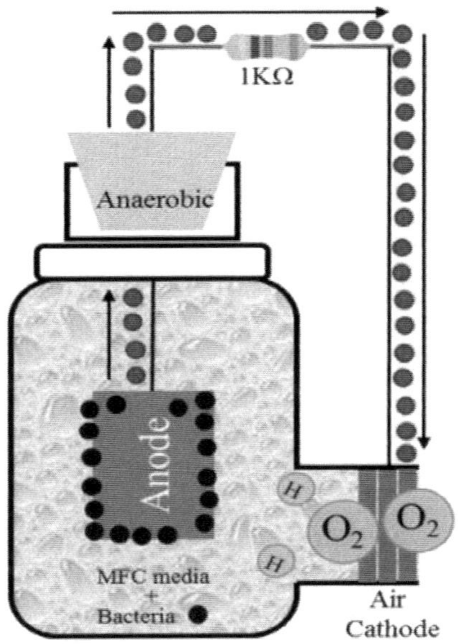

Figure 1. Single-chamber microbial fuel cell [50]

2.1. *Microorganisms*

Normally blended bacterial culture is utilized in SCMFC since a wide scope of substrates can be dealt with. Yet, there are a few reactors wherein single bacterial species are utilized [48]. Depending upon the exchange of created electrons from the substrate to anode by microbes, SCMFCs can be grouped into two sorts. The first is SCMFCs with a mediator, where outer chemical mixes are utilized for electron movement in the anode chamber and the subsequent one is, mediator-less MFCs where the mediator is missing [43,51]. The creation of power by Shewanella putrefaciens without the help of external acceptors of electrons was analyzed in an SCMFC with various kinds of anodes and alternative physical conditions by Park and Zeikus [22]. It was discovered that power creation was subject to anode organization, type of electron donor and concentration of cells. A most extreme current density of 10.2 mW/m^2 and 2.5 mA current was acquired with an anode of Mn^{4+} graphite, 3.9 g cell protein/ml cell concentration and sodium lactate of 20mM. Current generation by S. putrefaciens was upgraded 10-times when graphite anode was joined with Mn^{4+} which is an external mediator of electron. In an investigation, for the ongoing and consistent estimation of Cr^{6+} ions, three-stage SCMFC biosensor was created. Exiguobacterium aestuarii YC211 was inoculated in this sensor. At a liquid detention time of 2 mins, when Cr^{6+} concentration was increased from 5 mg/L to 30 mg/L, the voltage yield reduced. By providing three separate SCMFCs in a single arrangement, upto 90 mg/L concentration of Cr^{6+} can be estimated. As per the cutting edge sequencing analysis, Cr(VI) in real tannery wastewater can be precisely quantified by this biosensor with less fluctuations. The solid accomplishment of this sensor is due to the stable bacterial network present in the MFC. Thus, for ongoing and persistent Cr^{6+} estimation of wastewater from tanneries, the three-stage SCMFC biosensor can be used extensively [52]. An MFC was worked with a pure culture of Lactobacillus pentosus and a synthetically prepared wastewater which was similar to an actual dairy industry wastewater. The general exhibition was assessed via COD degradation percentage and the power output. With the operating conditions of 61 cm^2 anode area, 25 cm^2 area of Nafion, yeast extract of 50 mg/L and a batch cycle of 2 days, a most extreme density of power of 5.04 mW/m^2 was accomplished. Eventhough after trying several operating conditions, only the power output from the MFC gets affected but treatment capacity of L. pentosus remains unaffected. For different operating conditions, the COD degradation percentage of dairy industry wastewater were somewhere between 42% and 58% [53]. Further in a study, the performance of membraneless single chambered microbial energy units were concentrated under anaerobic condition using the synthetic glucose substrate to produce power. The biofilm of organism Bacillus firmus – NMBL-03 over plain graphite terminals was utilized as a biocatalyst. When worked with an outer resistance of 1000 Ω, the MFC created the most extreme power output of 0.88 mW/m^2 and a normal power output of 0.19 mW/m^2 for 500 hrs [54]. Microbial energy components recognized by various strains of species of Pseudomonas; MFC-I:Pseudomonas aeruginosa and MFC II:Pseudomonas fluorescens were examined by Ilamathi et al. [55]. From the results, it have been demonstrated when 0.5 mM of dye was used as substrate, MFC I delivered the greatest power output of 2.87 mW m^{-2} for mono-azo reactive color and 1.91 mW m^{-2} for di-azo reactive dye and MFC II delivered a power output of 1.9 mW m^{-2} for mono-azo reactive color and 1.1 mW m^{-2} for di-azo reactive dye over a time of 360 h. Decolorization ability of MFC I was 98% for mono-azo reactive color and 95% for di-azo reactive dye. The use of external mediator phenazine diminished color reduction. The bioadhesion ability of Pseudomonas aeruginosa species on the anode electrode decreased the MFC inside resistance. In 2017, Khater et al. researched the working of acetate feed membraneless SCMFC and phyiological portrayal of the biofilm present on the surface of anode was studied utilizing 16S rRNA analyzer and Scanning Electron Microscope (SEM). The working of SCMFC was explored utilizing an anode made of carbon paper and a cathode of 0.3 mg/cm^2 area carbon paper treated with Teflon. The most extreme open circuit voltage (OCV) was seen as 791 mV. This framework effectively uncovered a higher power output of 86 mW/m^2 at a constant current density of 354 mA/m^2 with 65% coulombic effectiveness. COD removal percentage was observed as 96%. SEM analysis of the anode electrode exposed the presence of thick layer of microorganisms on the surface. Sequencing results from 16S rRNA analyzer shown that the biofilm on anode surface consists of a blend of different species of microbes instead of single microbial species. The predominant phyla are Alpha Proteobacteria, Firmicutes, Gamma Proteobacteria, Actinobacteria, with ten prevailing microbial strains: Bacillus firmus, Bacillus safensis, Bacillus isronensis, Micrococcus endophyticus, Brevundimonas bullata, Pseudomonas putida, Planococcus citreus, Acinetobacter tandoii, Shewanella profunda and Shewanella xiamenensis [56]. Microbial community examination in an investigation by Dai et al. demonstrated that Proteobacteria (80.7%), Gammaproteobacteria (48.1%), and Dokdonella (29.5%) ruled at the phylum, class, and family levels, respectively, on the anodic biofilm. This investigation offers a possible choice for the treatment of unmanageable recalcitrant organics, azo dyes and sulfide pollutants utilizing single-chamber air cathode MFCs [57]. In another study, direct production of power from lignocellulosic monosaccharides biomass was inspected utilizing SCMFCs with air cathodes. Power was produced from all the sources of carbon tried, including three pentoses and two uronic acids, one aldonic corrosive and mainly six hexoses. Eventhough the transformation time changed from one hour to seventy hours, the blended culture of bacteria adjusted with all the tried sources of carbon. The higher density of power acquired from these tried sources varies from 1240 to 2770 mW/m^2 with density of current varies from 0.76 to 1.18 mA/cm^2. The most minimal power output was brought about by d-Mannose and the most elevated was produced by d-glucuronic corrosive. Coulombic proficiency ranges from twenty one to thirty seven percentage. The assessed most extreme voltage output varied somewhere in the range between 0.26 and 0.44 V. Chemical oxygen demand (COD) was reduced by more than 80% for all the tried sources of carbon [58]. In another study, Bacillus subtilis species was used in an SCMFC without air cathode under anaerobic conditions. Here 1 mM of glucose was taken as the source of carbon and 1 mM of nitrate was taken as the electron acceptor. When the nitrate concentration was decreased in the MFC, a diminished current was noted. Further addition of nitrate, recuperated the current output. This demonstrated the reliance of Bacillus subtilis bacterial species on the electron acceptor i.e., nitrate to effectively create power. At an external resistance of 220 Ω, a power output of 1.9 $\mu W/cm^2$ was accomplished. Cyclic voltammograms (CV) demonstrated mediators in the MFC directly contribute to electrons. Because of glucose fermentation, low coulombic effectiveness (CE) of 11% was ascribed primarily. These outcomes showed that power can be retrieved from wastewater containing nitrate [59]. Saccharomyces cerevisiae as a biocatalyst in an MFC anode chamber was assessed for waste degradation and power generation. This MFC was worked in a single compartment design without external mediators utilizing graphite without catalysts as anodes. Two organic loading rates, 910 g COD/m^3-day (OLR-I) and 1430 g COD/m^3-day (OLRI-II) were taken. Maximum output of current was seen at pH 6.0 [160 mA/m^2 for OLR-I and 282 mA/m^2 for OLR-II] than pH 5.0 (137 mA/m^2) and pH 7.0 (129 mA/m^2) [60]. Similarly, the ability of mixed microbial culture was assessed under various values of pH i.e., pH 6 (acidic), pH 7 (neutral) and pH 8 (alkaline) during synthetic wastewater purification in a mediatorless SCMFC. It was operated at room temperature of 29±20^{0}C. Here only the bio-culture was used as anodic biocatalyst and the electrodes were non-catalyzed graphite terminals. It was found that the degradation relies upon the pH input. Maximum density of current was seen at pH 6 [acidic; 186 mA/m^2] contrasted with pH 7 [neutral; 146 mA/m^2] and pH 8 [alkaline; 135 mA/m^2]. Despite what might be expected, substrate reduction was discovered to be successful at neutral conditions (pH 7). Acidophilic activity indicated higher CE and powerful transfer of electrons at moderately higher resistance contrasted with neutral and alkaline pH examined [61]. As of late, Vilela et al. led an investigation for using a nanocomposite made of polyelectrolyte dependent on bacterial cellulose (BC) and poly(4-styrene sulfonic corrosive) (PSSA) as a sustainable PEM for application in MFCs. The use of the individual culture of *Shewanella frigidimarina* and PSSA/BC nanocomposite layer in a lab-scale SCMFC yielded OCV of 436 mV and a higher power output of 2.42 mW/m^2. These results were found to be far more superior to the results accomplished with a commercial Nafion® membrane as a PEM for MFCs [62].

2.2. *Anode*

In the effective design of SCMFC, the efficiency of electrode materials plays a pivotal role. The best anode material must have a large surface area, good electrical conductivity, toughness and mechanical strength [63]. Traditionally, carbonaceous materials in various forms are the most widely used anode materials. Non-catalyzed plain graphite electrodes were also used as anode materials in SCMFC. A setup was demonstrated for the purification of synthetic and real chemical wastewater from various industries. Here graphite plates of 70 cm^2 surface area were used. In this setup, always the bottom layer of the anode is immersed in wastewater [64]. Anodes made of plain graphite plates without catalysts were also used for bioelectricity production and it was documented that a voltage of 379 mV and a current density 538 mA/m^2 was generated under aerobic metabolic function [65]. In a work, activated carbon fiber (ACF) felt anode was used in a membraneless, air-cathode single-chamber MFC. Experiments were also conducted after treating ACF anodes with nitric acid and ethylenediamine and the results were compared [66]. Zhang et al. [67] in 2011 adopted a carbon paper modified with mesoporous carbon (MC) as a novel anode and checked its performance in an SCMFC. He constructed this anode by the method of layer-by-layer self-assembly. Carbon-based multiwalled nanotubes were also employed as anode materials for the purification of wastewater from distillery industry. During this work, the efficiency of nanotubes was compared with an anode, the surface of which was coated with carbon nanopowder and plain graphite anode. It was found that nanotubes were found to be more effective among the anodes [68]. For the treatment of dairy industry wastewater, spiral anode made of graphite-modified stainless steel mesh was used. When compared with previous studies, this novel anode is more effective which is confirmed by the higher power density of 20.2W/m^3 produced by them [69]. Simple untreated carbon in the form of brush and cloth were used for the degradation of acetate, glucose, domestic wastewater, etc [70–72]. An extreme power density of 1444 mW/m^2 was obtained with acetate as the substrate. 1540 mW/m^2 was obtained when glucose was used as substrate and 464 mW/m^2 when actual wastewater from domestic areas was used as pollutant. Xu et al. 2019 used carbon brush as anode after 30 min heat treatment at 450^{0}C and used it to study the development of the microbial community [73]. In 2019, Geethanjali et al. [74] synthesized $NiWO_4$ and reduced graphene oxide (NWG) nanocomposite by the one-pot solvothermal method and used them as an anode in SCMFC for treating glucose. At optimum operational parameters, 1458 mW/m^2 of highest density of power was obtained when NWG modified anode was used and it was 8.5

times more than the power density obtained with unmodified carbon cloth. Apart from these carbon-based materials, some natural materials were also used as anode electrode material. One such example is the usage of "JinHao" as an anode. It is a brand of bamboo charcoal. This compound is made from bamboo. Bamboo's outer area consists of a compact protective layer, a corneum and a layer of gelatinous structure. This will remain on its surface even after carbonizing it. The preparation of this low-cost electrode is as follows. First, the bamboo charcoals were polished and immersed in 1 N HCl overnight and washed with deionized water. It was then immersed in 1 N NaOH for one full night. Finally, it was cleansed in deionized water for many times. This prepared electrode was tested in an SCMFC fed with artificial wastewater [75].

2.3. Cathode

From Figure 2, it is concluded that the overall cost of SCMFC mainly depends on electrode materials, particularly cathode. Hence, the selection of cathode material plays a prominent role.

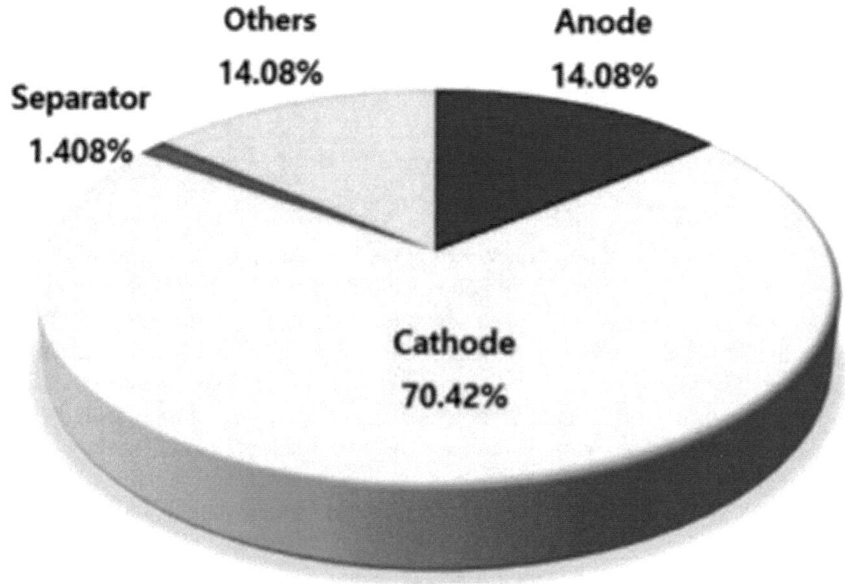

Figure 2. Cost analysis of SCMFC[76]

The cathode may be either present along with Proton Exchange Membrane (PEM) or membrane-less. In order to improve Oxidation Reduction Reaction (ORR), cathode may sometimes coated with a catalyst. In 2017, Jadhav et al. explored the efficiency of certain catalysts like palladium, manganese oxide, zirconium oxide and their composites in an SCMFC. Among the catalysts, Mn/Zr-Pd-C-based composite proved to be efficient since it generated a maximum power output of 1.28 W/m^3 [77]. Carbon electrodes coated with platinum are the most popular cathode materials. But they are costly. Thus, some low-cost cathode materials were prepared and employed in SCMFC. In a study by Chang et al. 2019 [78], an innovative biochar air cathode was utilized. This cathode was made from chips of Balsa wood after the method of pyrolysis at 800 °C. After pyrolysis, these chips become highly porous and also possess high ORR property. From the biochar air cathode SCMFC, a highest density of power of 200 mW/m^2 was reached. In these experiments, chips of biochar are used directly as air cathode without any binders, catalyst and GDLs. Similar work was carried out by Scicarria et al. [79] in which biochar was prepared from olive mill waste. Such SCMFCs achieved a highest power output of 271 ± 34 mW/m^2. It was then found that these results were 15 fold better than the power density obtained with commercial carbon cathodes. In another work by Allam et al. 2019 [80] Water Hyacinth Biochar

(WHB) was prepared by the process of pyrolysis at 900^{0}C. This WHB was placed in an air cathode SCMFC to act as ORR catalyst. The ability of production of power density by this catalyst was compared with another ORR Pt/C catalyst. The results showed that when compared with a costly Pt/C catalyst, this cheap WHB ORR catalyst generated two times higher power density. It demonstrated that in MFCs, cheap WHB is a good option as catalyst.

3. Applications of SCMFC in wastewater treatment

SCMFCs can be used in the purification of wastewater. Wastewater may be either synthetic which is prepared in the laboratory for research purposes or real which is produced in the industries during various processes.

3.1. Synthetic Wastewater

Khater et al. 2017 [56] designed a membraneless SCMFC and used acetate as a substrate. This MFC degraded 96% of acetate by generating 354 mA/m^2 constant current density, 86.1 mW/m^2 maximum power density and 65% coulombic efficiency. In the same year, another synthetic wastewater containing cefazolin sodium (CFZS) was prepared by Zhang et al. in 2018[81]. From the studies, it was found that SCMFC fed with CFZS generated electricity and removed the contaminant to a great extent. This SCMFC also had high tolerance with this contaminant which is usually present in hospital wastewater. When 100mg/L of CFZS was loaded in MFC, a highest output of power of 30.4±2.1 W/m^3 and a stable power density of 18.2±1.1 W/m^3 was obtained. Ethanolamine which has been used in the production of emulsifiers and detergents can be treated using SCMFC. This has been proved by An et al. in 2020 [82]. When 100 mg/L and 250 mg/L of ethanolamine were used in the SCMFC as substrates, considerable removal of COD and ammonia were attained. Synthetic human black water is also treated using SCMFC [83]. The mean power density over the external load was 153 ± 29 mW/m^2. Wen et al. 2011 [84] prepared synthetic wastewater with three compounds viz. glucose, penicillin and a combination of both penicillin and glucose. These wastewaters were fed into a single chamber MFC with open air-cathode and experiments were conducted by using them as fuel. From the outcomes, it was revealed that the simultaneous degradation of penicillin and the generation of electricity took place. In this demonstration, the awful factor is that these combination of penicillin and glucose produced electricity effectively. The highest power output for the mixture of 50 mg/L of penicillin and 1 g/L of glucose (101.2W/m^3) was 6 times more than the power output from 50 mg/L of penicillin (2.1 W/m^3) and 1 g/L of glucose (14.7 W/m^3) when used as separate fuels.

3.2. Real Wastewater

So far, several forms of wastewater collected from various industries have been tried in SCMFC and their removal efficiency and electricity production is studied. Applications of SCMFCs in actual wastewater treatment with their performances are listed in Table 1.

Table 1. Use of SCMFC in real wastewater treatment

Wastewater	SCMFC Configuration	Substrate concentration	Performance/Result	Reference
Swine wastewater	Single chamber air cathode MFC	Soluble COD - 8320 mg/L	261 mW/m^2 of highest power density was accomplished	[85]
Beer Brewery wastewater	Single chamber membrane-free MFC	COD - 2239 mg/L	Maximum power density of 483 mW/m^2	[86]
Swine wastewater	Single chamber MFC	Soluble COD - 8270 mg/L	Maximum power density of 228 mW/m^2 was generated and removed 84% of the organic matter	[87]
Confectionery wastewater	Single chambered air cathode MFC	COD - 1000 mg/L	Maximal power density of 373 mW/m^2 and 92% COD removal was	[88]

			obtained	
Starch processing wastewater	Single chamber air-cathode MFC with membranes	COD - 4852 mg/L	Maximum voltage output and power density of 490.8 mV and 239.4 mW/m^2 was reached	[89]
Composite wastewater from different industries, i.e., dyes, pesticides, pharmaceuticals and chemical process units.	Single chamber mediatorless open air cathode MFC	COD - 9.1 g/L	At acidic condition, a high density of current of 186.34 mA/m^2 was noted in comparison with the neutral condition where a density of current of 146 mA/m^2 and alkaline condition where a current density of 135.23 mA/m^2 was obtained.	[61]
Fermented wastewater	Single chamber MFCs with air cathodes	COD – 1.92g/L	At the organic loading rate (OLR) of 1.92 g/L/d, real fermented wastewater produced a power density of 1884 mW/m^3. When the OLR of fermented wastewater was increased to 3.84g/L/d, the value of power density increased to 2981 mW/m^3.	[90]
Different industrial wastewater	Single chamber open air cathode MFC	COD values: Paper industry - 600 mg/L Dairy industry - 700 mg/L Bakery - 651 mg/L Brewery - 661 mg/L	SCMFCs filled with wastewater from paper industry generated the maximum density of current of 125 mA/m^2 when compared with the SCMFCs filled with wastewater from dairy (25 mA/m^2), bakery and brewery industries (10 mA/m^2).	[91]
Dairy wastewater	Annular Single Chamber MFC	COD - 2500 to 5000 mg/L	Maximum power density of 20.2 W/m^3 and 91% COD removal was achieved	[69]
Olive mill wastewater	Single-chamber air cathode MFC	COD - 56.7 g/L	A maximum voltage of 381 mV was produced. The removal efficiencies of COD and phenolics were 65 % and 49 % respectively.	[92]
Distillery wastewater	Membrane-Less Air cathode single chamber MFC	COD - 125 to 3000 mg/L	Resultant current varies from 0.005 to 0.055 mA depending upon the concentration of distillery wastewater. 29.5-56.7% of soluble chemical oxygen demand (SCOD) and 35% of total solids was reduced.	[93]
Dye textile wastewater	Air-cathode single-chamber MFC	COD - 45.6 g/L	Maximum volumetric power density achieved was 123.2 ± 27.5 mW/m^3 .	[44]
Dye processing wastewater	Single chamber air cathode MFC	COD - 2100 mg/L	515 mW/m^2 of extreme power density was attained. Coulombic efficiency of 56 % and the high potency of total COD, soluble COD and TSS removal of about 85%, 73% and 68% respectively were reached.	[94]
Petroleum Refinery	Single chamber air cathode MFC	COD - 2150 mg/L	132 mW/m^2 of highest density of power was attained.	[95]

Wastewater				

4. Future Perspectives

From the review of earlier works, it is found that more importance should be given for the commercialization of SCMFCs. Power densities obtained from the studies already conducted were only a few thousands which is not enough for practical applications. These MFCs need to be stacked so that increased power output can be obtained. Research works need to be done on the use of low cost materials for the fabrication of MFC chamber and for the preparation of electrodes.

5. Conclusion

Microbial Fuel Cell is a novel technique used to purify wastewater and to generate electricity. It eliminates the wastewater which is a major pollutant of our environment. Also, the growing energy demand of the world is reduced by the production of electrical energy. This review paper emphasizes Single Chamber Microbial Fuel Cell, a type of MFC. The components, construction and working of SCMFC are discussed. The application of these MFCs for the treatment of synthetic and real wastewater is stated in this paper. The performance of these MFCs based on percentage of COD removal, power density, current density, coloumbic efficiency and voltage output is also mentioned.

References

[1] Yan P, Shi H-X, Chen Y-P, Gao X, Fang F and Guo J-S 2020 Optimization of recovery and utilization pathway of chemical energy from wastewater pollutants by a net-zero energy wastewater treatment model *Renew. Sust. Energ. Rev.* **133** 110160

[2] Sunitha A P, Sandeep K, Praveen P and Saji K J 2020 Upconverting carbon quantum dots: An eco-friendly material for light energy harvesting and bio-imaging *Mater. Today: Proc.*

[3] Yang W, Li J, Ye D, Zhang L, Zhu X and Liao Q 2016 A hybrid microbial fuel cell stack based on single and double chamber microbial fuel cells for self-sustaining pH control *J. Power Sources* **306** 685–91

[4] Chen Y, Kuang Y, Shi D, Hou M, Chen X, Jiang L, Gao J, Zhang L, He Y and Wong C-P 2020 A triboelectric nanogenerator design for harvesting environmental mechanical energy from water mist *Nano Energy* **73** 104765

[5] Abedi H, Abbassi A M and Capozziello S 2019 Cosmological perturbations in gravitational energy–momentum complex *Ann. Phys.* **405** 54–68

[6] Asaithambi P, Govindarajan R, Busier Yesuf M, Selvakumar P and Alemayehu E 2020 Enhanced treatment of landfill leachate wastewater using sono(US)-ozone(O3)–electrocoagulation(EC) process: role of process parameters on color, COD and electrical energy consumption *Process Saf. Environ. Prot.* **142** 212–8

[7] Choi J, Jung I and Kang C-Y 2019 A brief review of sound energy harvesting *Nano Energy* **56** 169–83

[8] Saidi K and Omri A 2020 Reducing CO_2 emissions in OECD countries: Do renewable and nuclear energy matter? *Prog. Nucl. Energy* **126** 103425

[9] Abdelaziz E A, Saidur R and Mekhilef S 2011 A review on energy saving strategies in industrial sector *Renew. Sust. Energ. Rev.* **15** 150–68

[10] Poudyal R, Loskot P, Nepal R, Parajuli R and Khadka S K 2019 Mitigating the current energy crisis in Nepal with renewable energy sources *Renew. Sust. Energ. Rev.* **116** 109388

[11] Chaurasiya P K, Warudkar V and Ahmed S 2019 Wind energy development and policy in India: A review *Energy Strategy Rev.* **24** 342–57

[12] Enamala M K, Enamala S, Chavali M, Donepudi J, Yadavalli R, Kolapalli B, Aradhyula T V, Velpuri J and Kuppam C 2018 Production of biofuels from microalgae - A review on cultivation, harvesting, lipid extraction, and numerous applications of microalgae *Renew. Sust. Energ. Rev.*_ **94** 49–68

[13] Kothari R, Vashishtha A, Singh H M, Pathak V V, Tyagi V V, Yadav B C, Ashokkumar V and Singh D P 2020 Assessment of Indian bioenergy policy for sustainable environment and its impact for rural India: Strategic implementation and challenges *Environ. Technol. Innov.* **20** 101078

[14] Kwan T H, Katsushi F, Shen Y, Yin S, Zhang Y, Kase K and Yao Q 2020 Comprehensive review of integrating fuel cells to other energy systems for enhanced performance and enabling polygeneration *Renew. Sust. Energ. Rev.* **128** 109897

[15] Sharma Y C and Singh V 2017 Microalgal biodiesel: A possible solution for India's energy security *Renew. Sust. Energ. Rev.* **67** 72–88

[16] Ghadge S V and Raheman H 2005 Biodiesel production from mahua (Madhuca indica) oil having high free fatty acids *Biomass Bioenergy* **28** 601–5

[17] Zheng Y H, Li Z F, Feng S F, Lucas M, Wu G L, Li Y, Li C H and Jiang G M 2010 Biomass energy utilization in rural areas may contribute to alleviating energy crisis and global warming: A case study in a typical agro-village of

Shandong, China *Renew. Sust. Energ. Rev.* **14** 3132–9

[18] You P Y and Kamarudin S K 2017 Recent progress of carbonaceous materials in fuel cell applications: An overview *Chem. Eng. J.* **309** 489–502

[19] Mekhilef S, Saidur R and Safari A 2012 Comparative study of different fuel cell technologies *Renew. Sust. Energ. Rev.* **16** 981–9

[20] Patel R and Deb D 2018 Parametrized control-oriented mathematical model and adaptive backstepping control of a single chamber single population microbial fuel cell *J. Power Sources* **396** 599–605

[21] Kwon C H, Ko Y, Shin D, Kwon M, Park J, Bae W K, Lee S W and Cho J 2018 High-power hybrid biofuel cells using layer-by-layer assembled glucose oxidase-coated metallic cotton fibers *Nat. Commun.* **9** 4479

[22] Park D H and Zeikus J G 2002 Impact of electrode composition on electricity generation in a single-compartment fuel cell using Shewanella putrefaciens *Appl. Microbiol. Biotechnol.* **59** 58–61

[23] Nasar A and Perveen R 2019 Applications of enzymatic biofuel cells in bioelectronic devices – A review *Int. J. Hydrog. Energy* **44** 15287–312

[24] Xiao B, Yang F and Liu J 2011 Enhancing simultaneous electricity production and reduction of sewage sludge in two-chamber MFC by aerobic sludge digestion and sludge pretreatments *J. Hazard. Mater.* **189** 444–9

[25] Grattieri M and Minteer S D 2018 Microbial fuel cells in saline and hypersaline environments: Advancements, challenges and future perspectives *Bioelectrochemistry* **120** 127–37

[26] Choudhury P, Ray R N, Bandyopadhyay T K, Basak B, Muthuraj M and Bhunia B 2020 Process engineering for stable power recovery from dairy wastewater using microbial fuel cell *Int. J. Hydrog. Energy*

[27] Foad Marashi S K and Kariminia H-R 2015 Performance of a single chamber microbial fuel cell at different organic loads and pH values using purified terephthalic acid wastewater *J. Environ. Health Sci. Eng.* **13** 27

[28] Yang N, Liu H, Jin X, Li D and Zhan G 2020 One-pot degradation of urine wastewater by combining simultaneous halophilic nitrification and aerobic denitrification in air-exposed biocathode microbial fuel cells (AEB-MFCs) *Sci. Total Environ.* **748** 141379

[29] Mohammed A J and Ismail Z Z 2018 Slaughterhouse wastewater biotreatment associated with bioelectricity generation and nitrogen recovery in hybrid system of microbial fuel cell with aerobic and anoxic bioreactors *Ecol. Eng.* **125** 119–30

[30] Lai M-F, Lou C-W and Lin J-H 2018 Improve 3D electrode materials performance on electricity generation from livestock wastewater in microbial fuel cell *Int. J. Hydrog. Energy* **43** 11520–9

[31] Zou H and Wang Y 2017 Azo dyes wastewater treatment and simultaneous electricity generation in a novel process of electrolysis cell combined with microbial fuel cell *Bioresour. Technol.* **235** 167–75

[32] Do M H, Ngo H H, Guo W, Chang S W, Nguyen D D, Liu Y, Varjani S and Kumar M 2020 Microbial fuel cell-based biosensor for online monitoring wastewater quality: A critical review *Sci. Total Environ.* **712** 135612

[33] Lee D-J, Chang J-S and Lai J-Y 2015 Microalgae-microbial fuel cell: A mini review *Bioresour. Technol.* **198** 891–5

[34] Gude V G 2016 Wastewater treatment in microbial fuel cells--an overview *J. Clean. Prod.* **122** 287–307

[35] Du Z, Li H and Gu T 2007 A state of the art review on microbial fuel cells: A promising technology for wastewater treatment and bioenergy *Biotechnol. Adv.* **25** 464–82

[36] Slate A J, Whitehead K A, Brownson D A C and Banks C E 2019 Microbial fuel cells: An overview of current technology *Renew. Sust. Energ. Rev.* **101** 60–81

[37] Antolini E 2019 Photoelectrocatalytic fuel cells and photoelectrode microbial fuel cells for wastewater treatment and power generation *J. Environ. Chem. Eng.* **7** 103241

[38] Cheng S, Liu H and Logan B E 2006 Power densities using different cathode catalysts (Pt and CoTMPP) and polymer binders (nafion and PTFE) in single chamber microbial fuel cells *Environ. Sci. Technol.* **40** 364–9

[39] Xia C, Zhang D, Pedrycz W, Zhu Y and Guo Y 2018 Models for Microbial Fuel Cells: A critical review *J. Power Sources* **373** 119–31

[40] Sonawane J M, Yadav A, Ghosh P C and Adeloju S B 2017 Recent advances in the development and utilization of modern anode materials for high performance microbial fuel cells *Biosens. Bioelectron.* **90** 558–76

[41] Choudhury P, Uday U S P, Mahata N, Nath Tiwari O, Narayan Ray R, Kanti Bandyopadhyay T and Bhunia B 2017 Performance improvement of microbial fuel cells for waste water treatment along with value addition: A review on past achievements and recent perspectives *Renew. Sust. Energ. Rev.* **79** 372–89

[42] Li M, Zhou M, Tian X, Tan C, McDaniel C T, Hassett D J and Gu T 2018 Microbial fuel cell (MFC) power performance improvement through enhanced microbial electrogenicity *Biotechnol. Adv.* **36** 1316–27

[43] Do M H, Ngo H H, Guo W S, Liu Y, Chang S W, Nguyen D D, Nghiem L D and Ni B J 2018 Challenges in the application of microbial fuel cells to wastewater treatment and energy production: A mini review *Sci. Total Environ.* **639** 910–20

[44] Logroño W, Pérez M, Urquizo G, Kadier A, Echeverría M, Recalde C and Rákhely G 2017 Single chamber microbial fuel cell (SCMFC) with a cathodic microalgal biofilm: A preliminary assessment of the generation of bioelectricity and biodegradation of real dye textile wastewater *Chemosphere* **176** 378–88

[45] Ghasemi M, Shahgaldi S, Ismail M, Kim B H, Yaakob Z and Wan Daud W R 2011 Activated carbon nanofibers as an alternative cathode catalyst to platinum in a two-chamber microbial fuel cell *Int. J. Hydrog. Energy* **36** 13746–52

[46] Ali A E-H, Gomaa O M, Fathey R, Kareem H A E and Zaid M A 2015 Optimization of double chamber microbial fuel cell for domestic wastewater treatment and electricity production *J. Fuel Chem. Technol.* **43** 1092–9

[47] Daud S M, Daud W R W, Kim B H, Somalu M R, Bakar M H A, Muchtar A, Jahim J, Lim S S and Chang I S 2018

Comparison of performance and ionic concentration gradient of two-chamber microbial fuel cell using ceramic membrane (CM) and cation exchange membrane (CEM) as separators *Electrochim. Acta* **259** 365–76
[48] Das S and Mangwani N 2010 Recent developments in microbial fuel cells: a review *J. Sci. Ind. Res.* **69** 727–31
[49] Vogl A, Bischof F and Wichern M 2016 Increase life time and performance of Microbial Fuel Cells by limiting excess oxygen to the cathodes *Biochem. Eng. J.* **106** 139–46
[50] Flimban S G A, Ismail I M I, Kim T and Oh S-E 2019 Overview of Recent Advancements in the Microbial Fuel Cell from Fundamentals to Applications: Design, Major Elements, and Scalability *Energies* **12** 3390
[51] Rahimnejad M, Adhami A, Darvari S, Zirepour A and Oh S-E 2015 Microbial fuel cell as new technology for bioelectricity generation: A review *Alex. Eng. J.* **54** 745–56
[52] Wu L-C, Wang G-H, Tsai T-H, Lo S-Y, Cheng C-Y and Chung Y-C 2019 Three-Stage Single-Chambered Microbial Fuel Cell Biosensor Inoculated with Exiguobacterium aestuarii YC211 for Continuous Chromium (VI) Measurement *Sensors* **19**
[53] Vilas Boas J, Oliveira V B, Marcon L R C, Simões M and Pinto A M F R 2019 Optimization of a single chamber microbial fuel cell using Lactobacillus pentosus: Influence of design and operating parameters *Sci. Total Environ.* **648** 263–70
[54] Singh S, Dwivedi C and Pandey A 2016 Electricity generation in membrane-less single chambered microbial fuel cell *Proc. Int. Conf. Control Comput. Commun. Mater. (ICCCCM)* pp 1-4
[55] Ilamathi R, Merline Sheela A and Nagendra Gandhi N 2019 Comparative evaluation of Pseudomonas species in single chamber microbial fuel cell with manganese coated cathode for reactive azo dye removal *Int. Biodeterior. Biodegradation* **144** 104744
[56] Khater D Z, El-Khatib K M and Hassan H M 2017 Microbial diversity structure in acetate single chamber microbial fuel cell for electricity generation *J. Genet. Eng. Biotechnol.* **15** 127–37
[57] Dai Q, Zhang S, Liu H, Huang J and Li L 2020 Sulfide-mediated azo dye degradation and microbial community analysis in a single-chamber air cathode microbial fuel cell *Bioelectrochemistry* **131** 107349
[58] Catal T, Li K, Bermek H and Liu H 2008 Electricity production from twelve monosaccharides using microbial fuel cells *J. Power Sources* **175** 196–200
[59] Nimje V R, Chen C-C, Chen H-R, Chen C-Y, Tseng M-J, Cheng K-C, Shih R-C and Chang Y-F 2012 A single-chamber microbial fuel cell without an air cathode *Int. J. Mol. Sci.* **13** 3933–48
[60] Raghavulu S V, Goud R K, Sarma P N and Mohan S V 2011 Saccharomyces cerevisiae as anodic biocatalyst for power generation in biofuel cell: influence of redox condition and substrate load *Bioresour. Technol.* **102** 2751–7
[61] Veer Raghavulu S, Venkata Mohan S, Venkateswar Reddy M, Mohanakrishna G and Sarma P N 2009 Behavior of single chambered mediatorless microbial fuel cell (MFC) at acidophilic, neutral and alkaline microenvironments during chemical wastewater treatment *Int. J. Hydrog. Energy* **34** 7547–54
[62] Vilela C, Cordeiro D M, Boas J V, Barbosa P, Nolasco M, Vaz P D, Rudić S, Ribeiro-Claro P, Silvestre A J D, Oliveira V B, Pinto A M F R, Figueiredo F M L and Freire C S R 2020 Poly(4-styrene sulfonic acid)/bacterial cellulose membranes: Electrochemical performance in a single-chamber microbial fuel cell *Bioresou. Technol. Reports* **9** 100376
[63] Logan B E, Hamelers B, Rozendal R, Schröder U, Keller J, Freguia S, Aelterman P, Verstraete W and Rabaey K 2006 Microbial fuel cells: methodology and technology *Environ. Sci. Technol.* **40** 5181–92
[64] Venkata Mohan S, Veer Raghavulu S and Sarma P N 2008 Influence of anodic biofilm growth on bioelectricity production in single chambered mediatorless microbial fuel cell using mixed anaerobic consortia *Biosens. Bioelectron.* **24** 41–7
[65] Venkata Mohan S, Mohanakrishna G and Sarma P N 2008 Effect of anodic metabolic function on bioelectricity generation and substrate degradation in single chambered microbial fuel cell *Environ. Sci. Technol.* **42** 8088–94
[66] Zhu N, Chen X, Zhang T, Wu P, Li P and Wu J 2011 Improved performance of membrane free single-chamber air-cathode microbial fuel cells with nitric acid and ethylenediamine surface modified activated carbon fiber felt anodes *Bioresour. Technol.* **102** 422–6
[67] Zhang Y, Sun J, Hou B and Hu Y 2011 Performance improvement of air-cathode single-chamber microbial fuel cell using a mesoporous carbon modified anode *J. Power Sources* **196** 7458–64
[68] Mohanakrishna G, Krishna Mohan S and Venkata Mohan S 2012 Carbon based nanotubes and nanopowder as impregnated electrode structures for enhanced power generation: Evaluation with real field wastewater *Appl. Energy* **95** 31–7
[69] Mardanpour M M, Nasr Esfahany M, Behzad T and Sedaqatvand R 2012 Single chamber microbial fuel cell with spiral anode for dairy wastewater treatment *Biosens. Bioelectron.* **38** 264–9
[70] Cristiani P, Carvalho M L, Guerrini E, Daghio M, Santoro C and Li B 2013 Cathodic and anodic biofilms in Single Chamber Microbial Fuel Cells *Bioelectrochemistry* **92** 6–13
[71] Daghio M, Gandolfi I, Bestetti G, Franzetti A, Guerrini E and Cristiani P 2015 Anodic and cathodic microbial communities in single chamber microbial fuel cells *N. Biotechnol.* **32** 79–84
[72] Cheng S, Liu H and Logan B E 2006 Increased power generation in a continuous flow MFC with advective flow through the porous anode and reduced electrode spacing *Environ. Sci. Technol.* **40** 2426–32
[73] Xu G, Zheng X, Lu Y, Liu G, Luo H, Li X, Zhang R and Jin S 2019 Development of microbial community within the cathodic biofilm of single-chamber air-cathode microbial fuel cell *Sci. Total Environ.* **665** 641–8
[74] Geetanjali, Rani R, Sharma D and Kumar S 2019 Optimization of operating conditions of miniaturize single chambered microbial fuel cell using NiWO4/graphene oxide modified anode for performance improvement and microbial

communities dynamics *Bioresour. Technol.* **285** 121337
[75] Yang S, Jia B and Liu H 2009 Effects of the Pt loading side and cathode-biofilm on the performance of a membrane-less and single-chamber microbial fuel cell *Bioresour. Technol.* **100** 1197–202
[76] Palanisamy G, Jung H-Y, Sadhasivam T, Kurkuri M D, Kim S C and Roh S-H 2019 A comprehensive review on microbial fuel cell technologies: Processes, utilization, and advanced developments in electrodes and membranes *J. Clean. Prod.* **221** 598–621
[77] Jadhav D A, Deshpande P A and Ghangrekar M M 2017 Enhancing the performance of single-chambered microbial fuel cell using manganese/palladium and zirconium/palladium composite cathode catalysts *Bioresour. Technol.* **238** 568–74
[78] Chang H-C, Gustave W, Yuan Z-F, Xiào Y and Chen Z 2020 One-step fabrication of binder-free air cathode for microbial fuel cells by using balsa wood biochar *Environ. Technol. Innov.* **18** 100615
[79] Pepè Sciarria T, de Oliveira M A C, Mecheri B, D'Epifanio A, Goldfarb J L and Adani F 2020 Metal-free activated biochar as an oxygen reduction reaction catalyst in single chamber microbial fuel cells *J. Power Sources* **462** 228183
[80] Allam F, Elnouby M, El-Khatib K M, El-Badan D E and Sabry S A 2020 Water hyacinth (Eichhornia crassipes) biochar as an alternative cathode electrocatalyst in an air-cathode single chamber microbial fuel cell *Int. J. Hydrog. Energy* **45** 5911–27
[81] Zhang E, Yu Q, Zhai W, Wang F and Scott K 2018 High tolerance of and removal of cefazolin sodium in single-chamber microbial fuel cells operation *Bioresour. Technol.* **249** 76–81
[82] An B-M, Seo S-J, Hidayat S and Park J-Y 2020 Treatment of ethanolamine and electricity generation using a scaled-up single-chamber microbial fuel cell *J. Ind. Eng. Chem.* **81** 1–6
[83] Vogl A, Bischof F and Wichern M 2016 Single chamber microbial fuel cells for high strength wastewater and blackwater treatment—A comparison of idealized wastewater, synthetic human blackwater, and diluted pig manure *Biochem. Eng. J.* **115** 64–71
[84] Wen Q, Kong F, Zheng H, Cao D, Ren Y and Yin J 2011 Electricity generation from synthetic penicillin wastewater in an air-cathode single chamber microbial fuel cell *Chem. Eng. J.* **168** 572–6
[85] Min B, Kim J, Oh S, Regan J M and Logan B E 2005 Electricity generation from swine wastewater using microbial fuel cells *Water Res.* **39** 4961–8
[86] Wang X, Feng Y J and Lee H 2008 Electricity production from beer brewery wastewater using single chamber microbial fuel cell *Water Sci. Technol.* **57** 1117–21
[87] Kim J R, Dec J, Bruns M A and Logan B E 2008 Removal of odors from Swine wastewater by using microbial fuel cells *Appl. Environ. Microbiol.* **74** 2540–3
[88] Sun J, Hu Y, Bi Z and Cao Y 2009 Improved performance of air-cathode single-chamber microbial fuel cell for wastewater treatment using microfiltration membranes and multiple sludge inoculation *J. Power Sources* **187** 471–9
[89] Lu N, Zhou S-G, Zhuang L, Zhang J-T and Ni J-R 2009 Electricity generation from starch processing wastewater using microbial fuel cell technology *Biochem. Eng. J.* **43** 246–51
[90] Nam J-Y, Kim H-W, Lim K-H and Shin H-S 2010 Effects of organic loading rates on the continuous electricity generation from fermented wastewater using a single-chamber microbial fuel cell *Bioresour. Technol.* **101 Suppl 1** S33–7
[91] Velasquez-Orta S B, Head I M, Curtis T P and Scott K 2011 Factors affecting current production in microbial fuel cells using different industrial wastewaters *Bioresour. Technol.* **102** 5105–12
[92] Bermek H, Catal T, Akan S S, Ulutaş M S, Kumru M, Özgüven M, Liu H, Özçelik B and Akarsubaşı A T 2014 Olive mill wastewater treatment in single-chamber air-cathode microbial fuel cells *World J. Microbiol. Biotechnol.* **30** 1177–85
[93] Tanikkul P and Pisutpaisal N 2015 Performance of A Membrane-Less Air-Cathode Single Chamber Microbial Fuel Cell in Electricity Generation from Distillery Wastewater *Energy Procedia* **79** 646–50
[94] Karuppiah T, Pugazhendi A, Subramanian S, Jamal M T and Jeyakumar R B 2018 Deriving electricity from dye processing wastewater using single chamber microbial fuel cell with carbon brush anode and platinum nano coated air cathode *3 Biotech* **8** 437
[95] Mohanakrishna G, Abu-Reesh I M, Kondaveeti S, Al-Raoush R I and He Z 2018 Enhanced treatment of petroleum refinery wastewater by short-term applied voltage in single chamber microbial fuel cell *Bioresour. Technol.* **253** 16–21

An Efficient Power Theft Detection Using Mean-Shift Clustering and Deep Learning in Smart Grid

G. Johncy[1], A.AnishaFelise[2]

[1] Assistant Professor, St.Xavier's Catholic College of Engineering, Nagercoil, India.

[2] PG Student, St. Xavier's Catholic College of Engineering, Nagercoil, India.

Abstract – Energy theft constitutes a major concern for the utility operators in these modern smart homes. The task of detecting and reducing the energy losses has been highly challenging due to insufficient inspection techniques. Energy distribution is comprised of Technical and Non- Technical Losses (NTL). Energy theft generates a major share of Non-Technical Losses which also prompts budgetary misfortunes for the service organizations. The data in the modern smart meters are transmitted in wireless mode. Therefore the smart homes are vulnerable to energy theft. Many new technologies have been adopted to resolve the issue of energy theft in Advance Metering Infrastructure (AMI) in Smart Grids. Consumption pattern must be derived to identify illegal energy consumers. Computational method has been derived to analyze and identify energy consumption patterns based on data mining techniques. Machine learning technique improves the got client energy utilization readings and guides them on contrasting irregularities in use. Deep Learning method as Convolution Neural Network is implemented on the activity of order methods on client energy utilization and illegal use of electricity and the amount of consumption by energy theft.

Keywords: Smart Grid, Non-Technical Loss, Mean shift Clustering, Convolutional Neural Network.

1. Introduction

The electrical grid is comprised of segments, for example, transmission lines, substation furthermore, transformers where the energy is delivered from the power plant to the individual or a business [5][6]. The modern Grid is provided with the digitized two way communication between the consumers and the energy suppliers. A substantial energy loss occurs during the energy distributions. The loss may be either Technical Loss or Non- Technical Loss. Technical Losses are caused by power dissipation in the component. Non- Technical Losses are more specifically due to the energy theft. Generally, techniques for power utilization are i.e consuming all power from matrix legally; ii. Pilfering all power from network illegally; iii. Pilfering a bit of family unit power and expending the rest legitimately.[1][2]

Compared to the traditional meters which can be physically tampered, smart meters may not be tampered physically but the energy theft can happen due to cyber-attacks. An efficient way to resolve this issue to analyze the consumption pattern of the energy in the grid [4].

As of late, digital altering has become an approach to malignantly change the utilization information examples of the smart meter which leads to financial loss to the utility companies. Despite the fact that such power misfortune is brought about by few customers, it diminishes the benefit and energy

IOP Conf. Series: Materials Science and Engineering 983 (2020) 012003 doi:10.1088/1757-899X/983/1/012003

effectiveness of the utility organizations, making issues, for example, load-shedding, mechanical routine hampering, and expansion. "NTL costs a huge amount of money for both developed and developing countries such as USA, UK, Brazil, Malaysia, and India. For example, these losses in the power sectors of India and Brazil cost approximately US $44.5 and US $3 billion per year respectively" [6].

To mitigate this issue, a few procedures have been actualized through productive examination to recognize and subside the theft issue. Analysts arranged energy theft location frameworks into three essential gatherings, for example, state-based, game hypothesis based and order based identification. The step by step power usage examples of clients can be analyzed by utilizing AI calculations to set up characterization models, which incorporate choice trees, arbitrary backwoods (RF), uphold vector machines (SVM), neural organizations (NN, etc.[7][8].
The intermittent portrayal of intensity utilization since the consumption of power by the users is more huge in the field of ETD measure of information [9]. Due to the amount of data handled by the service organizations' AI based arrangement has accomplished huge consideration in fraud data detection. Day by day utilization of information is applied to distinguish theft pattern which protects the data protection of the shoppers. Means Shift algorithm was applied to identify peculiarities and anomalies in the gathered information by bunching and Convolutional Neural Network (CNN) is applied to classify the user profile [10].

In this paper, we propose another technique by consolidating a profound learning model with conventional AI models to detect the electricity theft. In particular, we utilize the Convolutional Neural Network (CNN) with Mean shift clustering model to accomplish better exactness and versatility. The strategy has two stages containing complete four stages. In the initial step, we collected the smart meter data, and then preprocess the crude information of records to eliminate insignificant elements. Also, Principal Component Analysis and Mean Shift clustering are applied to find the trustworthiness of customers. Thirdly, a CNN is prepared with these subsets to identify the Energy burglary. Finally, validation is performed on the testing subsets.

2.Related work

Power burglary location can be treated as unique case of anomaly identification or misrepresentation recognition. As an exceptional time series, electricity utilization records have pulled in a lot attention in the recent years. The everyday power utilization patterns of clients can be dissected by utilizing highlight based AI calculations to set up grouping models. Jindal et al. proposed [7] scheme dependent on decision tree and support vector machine is able to distinguish and find ongoing power theft at each level in power transmission and distribution. This plan lessens bogus positives generally and demonstrates its viability in the genuine situations. Results demonstrate that the plan distinguishes fake buyers with an exactness of 92.5% and a bogus positive rate as low as5.12%.

Depuru et al.[8] reported a neural organization (NN) model with a various leveled model for upgraded assessment of the order effectiveness and an encoding method can recognize illicit customers with better proficiency and quicker arrangement of information.

Zheng et al. [10] proposed that the irregular power use examples can be caught by wide and profound convolutional neural organization. Rajiv et al. proposed that Machine Learning calculation Gradient Boosting Theft Detector with include designing declines False Positive Rate and furthermore diminishes client information extra room just as preparing time [11].

Shuan et al.[12] reported a crossover convolutional neural organization irregular timberland model for automatic power robbery identification shows that the location model beats other methods as far as exactness and effectiveness.

Nazmul et al.[13] proposed an electricity theft detection system is CNN-based LSTM model for keen lattice information grouping on the grounds that the force utilization mark is time-arrangement information and an information pre-handling calculation was executed to process the missing occasions in the dataset, in view of the neighborhood esteems comparative with the missing information point. The engineered minority over-examining procedure was actualized to produce new information focuses so as to address the class awkwardness issue. Furthermore, the tally of power robbery clients was moderately low, which might have made the model to distinguish burglary clients with great exactness.

Zheng et al. examine the fundamental structure of the onlooker meters and the smart meters and a connection based identification strategy utilizing MIC is given to measure the relationship between the altered burden profiles and the NTL. CFSFDP-based strategy improves the location precision and soundness[14].Buzau et al. investigates the raw daily energy consumption history and incorporates non-consecutive information, for example, its contracted force or geological data. The results show that the data mining techniques to understand the problem. One strategy is the Maximum Information Coefficient (MIC), which can find the connections between the non-specialized misfortune (NTL) and a particular influence conduct of the buyer. MIC can be utilized to exactly recognize burglaries that seem ordinary in shapes [15].

Existing System

The past exploration takes a shot at electricity theft detection centers around the client power use profile information for theft identification. The specific are where there is dissimilarity in supplied power and billed power. All the customers belonging to that area are considered to be suspects. The drawback of the work discussed previously is that power theft identification has been carried out based on the assumption that the customers are suspected to be fraud. This case could identify the potential customer as the fraudulent customers. This provides the motivation for this research work to include the bogus customer power utilization information into the genuine force utilization information. The machine learning algorithms are used to analyze the data that literally cluster and deep learning algorithm is used classifies the Customer. The customer's data are discriminated as genuine and fraud based on their usage pattern.

A. System Model

This clever thought of Smart Energy Theft System for the shrewd home is more proficient and dependable contrasted with past techniques. Because of a non-meddling strategy for information assortment, the energy observing framework was actualized in a genuine house in Singapore. The gathered information incorporates Time arrangement information power utilization from a non-controlled genuine house environment

Keen Homes are made through execution of Internet of Things (IoT) and smart meters. So as to screen and control the Advanced Metering Infrastructure (AMI), Energy Management System (EMS) was a basic reconciliation of the framework foundation. Request Side Management System (DSMS) is incorporated as an element of EMS. Its usefulness centers principally around dealing with the interest reaction and burdens. It gathers the interest data to direct the ideal force use, for example, executing load-moving to empower the utilization of power markets during top and off-top hours. It permits clients to advantageously direct their shrewd machines inside the home territory by utilizing cell phones. Further developed and created frameworks could additionally break down the information gathered and settle on its own choice for the keen homes to work in a smart and energy-effective technique dependent on clients' utilization designs. Energy burglary has become a difficult issue in the keen matrix network. It has caused monstrous misfortunes for some nations that surpass billions of dollars.

3. Proposed system

Electricity theft is the criminal act of taking power through unlawful strategies without paying for the devoured power tax which brings about income misfortune for the utility supplier. The proposed model is shown in Figure 1.1

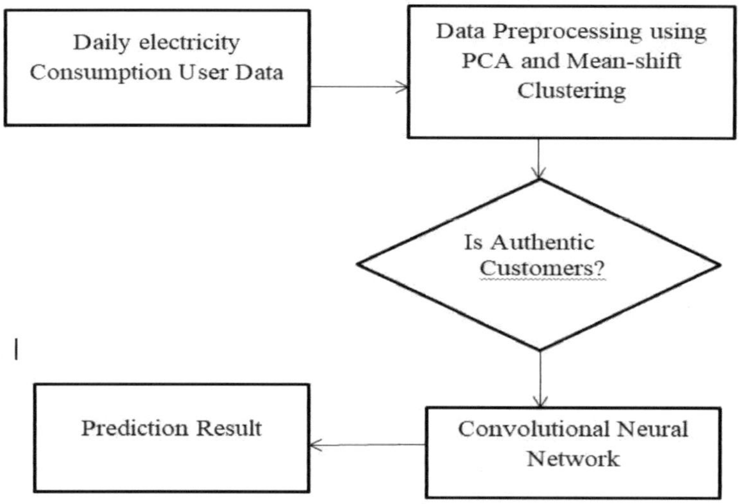

Figure 1.1. Proposed Model for Power theft identification

A. Data Preprocessing

The collected Smart meter data, the customer profile data of a specific area is chosen. All the customers of the selected region are genuine profiled. The percentage of trustworthiness of the customers is identified by using machine learning algorithm. For this purpose, unsupervised Mean Shift clustering algorithm is applied. The clustering is performed to select the customer's profile which is more genuine in their power usage pattern by grouping them in clusters. The customers are clustered based on their power usage readings obtained from their smart meter. For this purpose, a sample customer's smart meter reading for every 30 minutes in (KWH) for 28 days is considered.

B. Principal Component Analysis

At First, it needs to perform mean standardization at highlight scaling, so the highlights, for example, power utilization have zero mean and should have a practically identical scope of qualities using

$$\mu j = \frac{1}{m}\sum_{i=1}^{m} Xj(i)(1)$$

where$X_j$is the jth consumer data vector and the average electricity consumption of each consumer. Then we replace each (i) with((i)–μ_j).For datasets dimensionality reduction next step is to apply the technique principal component analysis which increasing interpretability but at the same time minimizing information loss . PCA obtains eigenvalues and eigenvectors which represent the characteristics and relationship of the data

C. Mean Shift Clustering

Mean shift algorithm is a non-parametric grouping technique which doesn't need prior data on the amount of cluster, and doesn't compel the state of the cluster. For every data point, mean shift algorithm characterizes a window around it and registers the mean of the information point. By then it shift the point of convergence of the window to the mean and rehashes the figuring till it combines.

Mean shift ALGORITHM
Input: Bandwidth h, profile function $g(x)$, and data matrix $X = [x_1, \dots . x_n], n \geq 2$
Output: Estimated cluster..
begin
for$(i = 1\ to\ n)$ do
Initialize: $j \leftarrow 1; y_j \leftarrow x_i$
The underlying center community y_1 is picked to be one of the watched information

repeat
Evaluate the mean shift vector

$$m(y_j) \leftarrow \frac{\sum_{i=1}^{n} x_i g\left(| | \frac{y_j - x_i}{h} | |^2 \right)}{\sum_{i=1}^{n} g\left(| | \frac{y_j - x_i}{h} | |^2 \right)} - y_j$$

Do the cluster center update

$$\widetilde{y_{j+1}} = m(y_j) + y_j$$

Find the nearest data point

$$y_{i+1} = arg \min_x \in \{x_1, \dots x_n\} \ | | \widetilde{y_{j+1}} - x | |$$

Until convergence happens,
$y_{i+\ 1} = y_j$ for some $j \in N$

$$y_i^* \leftarrow y_j$$

Consolidation the assessed cluster communities that are nearer than h
optional: Estimate cluster that pull in modest number of information focus end

Mean shift calculation ensure the assembly of the isolated arrangement. MS calculation appoints the processed MS Vector to the closet data point in the informational collection in every cycle. The quantity of emphases is exceptionally needy to the halting measures in the mean shift calculation however in proposed mean shift calculation we show that the quantity of cycles is constantly limited over the quantity of tests. In mean shift calculation, we need to consolidate the assessed group places that are nearer than the transmission capacity. We have to kill the assessed cluster community that draws in barely any number of focuses.

CNN Model

A CNN is a subclass of neural organization proposed, which is propelled by the working rule of utilizing the human visual cortex for object acknowledgment. CNNs were intended for recognizing objects, just as their classes, in a picture is appeared in Figure 1.2. CNN fluctuates from conventional AI calculations concerning incorporate element extraction, where CNN removes includes internationally through different stacked layers

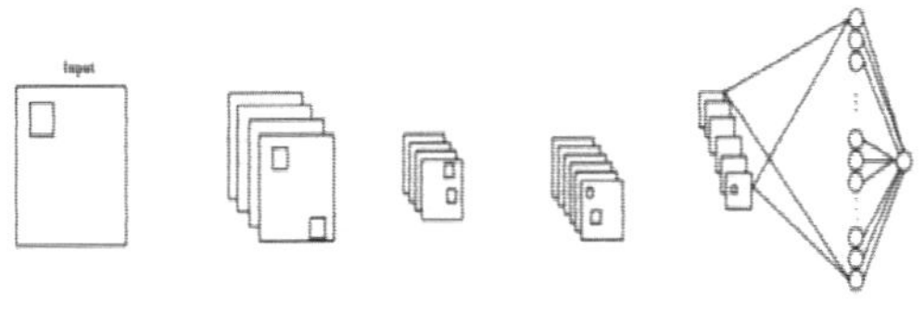

Figure 1.2 Convolutional Neural Networks

In, general CNN design involves a few convolution layers and pooling layers. These layers are trailed by one or more fully connected(FC) layers. The convolutional neural organization is the chief structure square of a Convolutional Neural Network. Convolution is a numerical activity that follows up on two sets of information.

$$Y=X * F\rightarrow y[i] = \sum_{j=-\propto}^{+\propto} x[i-j]F[j]$$

On account of CNNs, the two arrangements of data are the info information and a convolution channel, which is additionally called the portion. The convolution activity is performed by sliding the bit over the entire data, which delivers an element map. Practically speaking, various channels are used to play out different convolutions to create particular component maps. These element maps are at long last consolidated to define the last yield from the convolution layer. Initiation capacities are utilized after the convolution activity to acquaint non-linearity with the model. Distinctive enactment capacities, for example, straight capacity, sigmoid, and tanh are utilized, yet the corrected direct unit (ReLU) was utilized in the proposed CNN since it can prepare the model quicker and guarantee close worldwide weight improvement. The ReLU initiation work is characterized as follows:

$$f(x_i) = \max(0, x_i)$$

The pooling layer shows up close to the convolution layer. These layer down-examples each component guide to decrease their measurements, which thus diminishes over fitting and preparing time. The maximum pooling is generally utilized in CNNs which just chooses the most extreme incentive in the pooling window. The FC layer is fundamentally a completely associated fake neural organization. Basically, in a CNN, the convolution and pooling layers remove low-level highlights, for example, edges, lines, ears, eyes, and legs, and the completely associated layer plays out the order task dependent on these low-level highlights. The initiation work utilized in this last arrangement layer is normally a SoftMax work, which allocates a likelihood incentive to each class which includes to 1

4. Result and Discussion

The first step for the electricity theft detection is to input the consumer electricity data set. It contains the field like consumer no, date, area no, daily power consumption and hourly consumption. The customer profiles are assigned into four cluster 0,1,2 and 3. The usage pattern of finalized cluster head originated as the source for the anomaly customer detection. The customers profile which are close to the cluster head is identified as genuine profile. The genuine profiles are separated and their energy consumption data are taken for further investigation. The trustworthiness of the customer is verified by using the clustering algorithm is shown in Figure 1.3

```
[ 1.192e+00  1.291e+00]
[-5.960e-01  1.121e+00]
[ 1.100e+00 -9.008e-01]
[-6.400e-01  1.806e+00]
[ 1.388e+00 -1.954e-01]
[ 1.129e+00  6.183e-01]
[ 8.768e-01  8.348e-01]
[-5.978e-01  9.631e-01]
[ 6.066e-01 -1.137e+00]
[ 2.309e-01 -1.205e+00]
[ 9.176e-01 -1.393e+00]
[-7.686e-01 -1.071e+00]
[-5.840e-01 -1.553e+00]
[ 6.151e-01  3.590e-01]
[-1.259e+00  1.412e+00]
[ 6.869e-01 -1.534e-02]
[-1.626e+00  7.810e-01]
[-8.412e-01 -1.070e+00]
[ 8.609e-01 -1.182e+00]
[ 9.294e-01  1.257e+00]
[ 5.788e-01 -6.814e-01]
[-1.101e+00  1.304e+00]
[ 7.720e-01  8.352e-01]
[-1.423e-01  1.390e+00]]
assined clusters: [0, 0, 0, 1, 1, 0, 1, 2, 2, 3, 2, 0, 1, 0, 1, 3, 3, 3, 0, 0, 1,
2, 0, 1, 2, 2, 1, 1, 3, 2, 1, 3, 3, 2, 3, 0, 0, 0, 1, 2, 1, 3, 0, 1, 3, 0, 1, 3,
0, 3, 3, 0, 3, 3, 1, 0, 3, 2, 1, 1, 1, 2, 2, 1, 2, 3, 1, 1, 3, 3, 3, 0, 0, 3, 2,
2, 1, 0, 3, 2, 2, 1, 2, 1, 0, 1, 1, 3, 2, 1, 1, 3, 3, 3, 3, 3, 3, 0, 0, 2, 2, 3,
2, 2, 0, 1, 0, 2, 2, 1, 1, 1, 1, 3, 2, 2, 0, 1, 0, 3, 2, 0, 2, 1, 1, 3, 1, 2, 0,
0, 2, 0, 3, 1, 0, 0, 2, 2, 0, 0, 2, 2, 3, 2, 3, 0, 1, 3, 0, 2, 2, 2, 0, 3, 1, 3,
2, 0, 0, 2, 1, 1, 1, 2, 2, 0, 1, 2, 3, 2, 1, 2, 1, 2, 1, 1, 3, 3, 2, 3, 1, 3, 3,
0, 3, 2, 1, 2, 0, 3, 1, 0, 3, 2, 0, 1, 1, 1, 1, 0, 3, 0, 3, 2, 0, 0, 2, 0, 2, 2,
2, 0, 3, 3, 2, 3, 3, 3, 1, 3, 0, 0, 2, 0, 2, 0, 1, 0, 1, 0, 1, 3, 0, 3, 0, 0, 2,
0, 3, 0, 1, 2, 1, 2, 3, 3, 0, 1, 3, 1, 1, 0, 0, 2, 3, 1, 0, 1, 0, 3, 3, 0, 0, 2,
2, 3, 1, 1, 2, 2, 1, 0, 0, 2, 0, 2, 2, 3, 1, 3, 1, 2, 2, 3, 1, 1, 1, 0, 0, 2, 3,
1, 3, 0, 1, 2, 1, 3, 2, 3]
```

Figure 1.3. Clustering

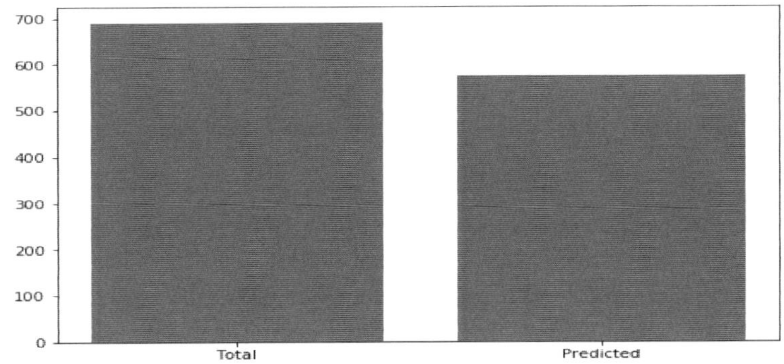

Figure 1.4.Classification

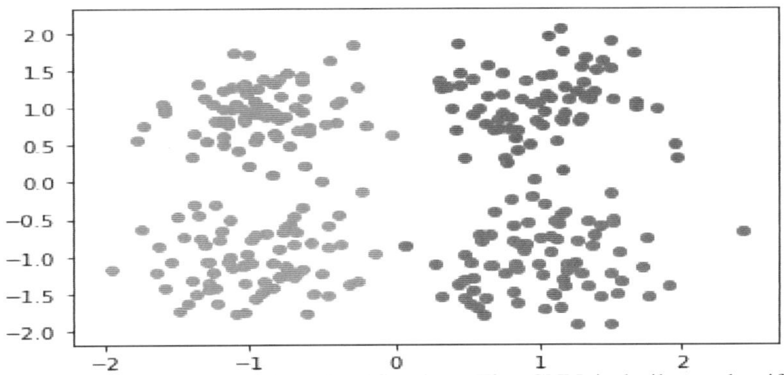

The clustered result is given to the input of CNN classification. The CNN is built to classify the customer's profile. The three types of bogus data along with the actual data are considered to train the network. After required number of iterations, the network is trained to predict any new customer profile to genuine or fraud shown in the Figure1.4. From the 2000 data sample 695 is detected as fraud, but by using CNN the predicted data is 574. Thus this model produces Accuracy rate of 85.9%, detection rate of 83%, False positive rate is 5% and it shown that the time taken to execute is 5ms/steps. Recall is 0.897% model F1 is 0.84.For CNN model if we increase the dataset it will produce more accurate result.

5. Conclusion

This project proposes an efficient way of using deep learning algorithm to address the power theft problem. In shrewd framework information examination framework, it is important to know the continuous power utilization information to estimate the specific future interest of power and plan in like manner. The identification of power burglary will likewise expand its help for load gauging that allows the utilities to precisely foresee the force interest for future explicit to singular client. The information produced through this analytics, increment information on client utilization design and the requirement of power for the future

Acknowledgement

G Johncy was graduated in Computer Science and Engineering in 2005 from Anna University-Chennai, India. In 2008, she received M.E degree in Computer Science and Engineering from Anna

University- Chennai, India. She is in teaching profession for the past Twelve years. Currently she is working as an Assistant Professor in the Department of Computer Science at St. Xavier's Catholic College of Engineering, India. Her areas of interest are Network Security, Image Processing and Smart grid Security. She is a Life member of ISTE.

A.Anisha Felise was graduated in Computer Science and Engineering in 2018 from Anna University-Chennai, India. In 2020, she received M.E degree in Computer Science and Engineering from Anna University- Chennai, India.

REFERENCES

[1] Depuru SS, Wang L, Devabhaktuni V, Green RC 2013 High performance computing for detection of electricity theft, *International Journal of Electrical Power & Energy Systems* **7** 21-30.
[2] Depuru SS, Wang L and Devabhaktuni V 2011 Electricity theft: Overview, issues, prevention and a smart meter based approach to control theft. *Energy policy* **39** 1007- 15.
[3] Li W, Logenthiran T, Phan VT and Woo WL 2019 A novel smart energy theft system (SETS) for IoT-based smart home. *IEEE Internet of Things Journal* **6** 5531-9.
[4] Glauner P, Boechat A, Dolberg L, State R, Bettinger F, Rangoni Y and Duarte D 2016 Large-scale detection of non-technical losses in imbalanced data sets. *IEEE Power & Energy Society Innovative Smart Grid Technologies Conference (ISGT)* 6 1-5.
[5] Depuru, Soma ShekaraSreenadh Reddy 2012 Modeling, detection, and prevention of electricity theft for enhanced performance and security of power grid.The University of Toledo.
[6] Prasad J, Samikannu R 2017 Overview, issues and prevention of energy theft in smart grids and virtual power plants in Indian context. *Energy Policy* **110** 365-74.
[7] Jindal A, Dua A, Kaur K, Singh M, Kumar N and Mishra S 2016 Decision tree and SVM-based data analytics for theft detection in smart grid. *IEEE Transactions on Industrial Informatics* **12** 1005-16.
[8] Depuru SS, Wang L, Devabhaktuni V and Nelapati P 2011 A hybrid neural network model and encoding technique for enhanced classification of energy consumption data. *In2011 IEEE Power and Energy Society General Meeting* **24** 1-8.
[9] Giani A, Bitar E, Garcia M, McQueen M, Khargonekar P and Poolla K. Smart grid data integrity attacks:characterizations and countermeasures π. *In2011 IEEE International Conference on Smart Grid Communications (SmartGridComm)* **17** 232-237).
[10] ZhengZ, YangY, NiuX,DaiHN and ZhouY 2017 Wide and deep convolutional neural networks for electricity-theft detection to secure smart grids. *IEEE Transactions on Industrial Informatics* **14** 1606-15.
[11]Punmiya R, Choe S 2019 Energy theft detection using gradient boosting theft detector with feature engineering-based preprocessing *IEEE Transactions on Smart Grid* **10** 2326-9.
[12]Li S, Han Y, Yao X, Yingchen S, Wang J and Zhao Q 2019 Electricity Theft Detection in Power Grids with Deep Learning and Random Forests *Journal of Electrical and Computer Engineering* Oct1 2019.
[13]Hasan M, Toma RN, Nahid AA, Islam MM and Kim JM 2019 Electricity theft detection in smart grid systems: a CNN-LSTM based approach *Energies* **12** 3310.
[14]Zheng K, Chen Q, Wang Y, Kang C and Xia Q 2018 A novel combined data-driven approach for electricity theft detection. *IEEE Transactions on Industrial Informatics* **15** 1809-19.
[15]Buzau MM, Tejedor-Aguilera J, Cruz- Romero P and Gomez-Exposito A 2019 Hybrid deep neural networks for detection of non-technical losses in electricity smart meters. *IEEE Transactions on Power Systems* **35** 1254-63.

An experimental study on the feasibility of using combination of industrial by-products and wastes as highway construction materials

S. Judes Sujatha[1*] J. Sahaya Ruben[2*], A.Sumil Kumaran[3]

[1*]*Department of Civil Engineering, University College of Engineering, Nagercoil, Tamil Nadu, India, Pin: 629 004. E-mail: judessuja@yahoo.com, Tel.: +91 9442054675, Fax: +91 4652 260511.*
[2]*Department of Civil Engineering, Rohini College of Engineering and Technology, Kanyakumari-629 401, Tamil Nadu, India, E-mail: rubenjsr1@gmail.com, . Tel:+91 9894176118*
[3]*Department of Civil Engineering, EASA College of Engineering and Technology, Coimbatore -641105, Tamil Nadu, India, Tel:+91 7418865896*
[*]*Corresponding Author*

Abstract. Use of industrial by-products and wastes as highway construction materials help achieving sustainable development. This paper presents and discusses the results of the experimental studies carried out to determine the effect of combination of different industrial by-products and wastes on unconfined compressive strength variation of the mixtures. The effect of different stabilizing agents such as cement and lime, and the effect of different curing periods such as 7-days, 14-days and 28-days on the strength variation are also examined.

Keywords. Highway Materials, Waste Foundry Sand, Fly Ash, Red Mud, Sustainability, Environment

1. INTRODUCTION

In light of achieving sustainable development, use of industrial by-products and wastes for construction of highways and village roads may be considered to be one of the effective ways to reduce carbon footprint. Since road construction consumes large volumes of materials, there is wide scope for using large volumes of industrial by-products and wastes that are normally disposed of on vacant land thereby preventing or reducing adverse effects on the environment. There are different industrial by-products and wastes that may be considered for use as partial or full replacement of conventional highway construction materials. Some of them are waste foundry sand (WFS), fly ash (FA), red mud (RM), blast furnace slag and cement kiln dust. When the quantity of some of these industrial by-products and wastes are insufficient, possibility of using combination of these industrial by-products and wastes may be considered. Nevertheless, studies on the effect of using combined industrial by-products and wastes on geotechnical strength parameters have to be carried out in order to make a conclusive statement. It is also to be noted that, combination of different wastes materials may also help achieving reduced disposal of wastes from different types of industries.

Sen and Mishra (2010) reviewed the possibilities of using the industrial wastes such as waste foundry sand, phospogypsum, fly ash, blast furnace slag, waste plastic bags and colliery sand for the construction of village roads. They reported that, use of these materials improves the engineering properties of the soil. Schroeder (1994) had listed fifteen waste materials that have been used in the past or being used successively at present for highway or embankment constructions.

Sybilski (2004) done the investigations to determine the possibilities of using different industrial wastes for highway construction. The author suggests that new guidelines and recommendations must be developed so as to effectively use the industrial wastes and by-products for highway construction.

Kirk (1998) showed that waste foundry sand from ferrous industries can provide engineering properties that are suitable for construction of highway embankment. The geotechnical test results that arrived indicate the strength and deformation performance of WFS as a structural fill is comparable to that of made using natural sand. Nevertheless, it is was noted that WFS fill cannot be considered to be freely draining and WFS material did not result in toxicity higher than that one can expect from using the natural sand. The author also pointed out that lack of decision, based on the scientific tools such as life-cycle or risk-based analysis methods form the barricades to reuse WFS for beneficial purposes. Mast and Fox (1998) studied the use of industrial waste foundry sand as a highway embankment material by conducting the field demonstration. Geotechnical concerns such as strength, deformation and hydraulic conductivity were evaluated in the highway project.

Goodhue et al. (2001) have noted that to facilitate beneficial reuse of WFS, typical strength parameters of the soil should be available to the designers. Towards providing the required design parameters, they have conducted the testing to characterize the strength of WFS and their interaction with geosynthetics. Tests such as multistage interface shear tests, direct shear tests, and pull-out tests were carried out in their study. The results of the tests indicate that WFS can be used in geotechnical construction effectively. Guney et al. (2006) have carried out laboratory experiments on soil-WFS mixtures to test their strength parameters and hence the suitability of the mixture to use as highway sub base materials. Cement and lime were used in their study as stabilization materials. The considered mixture proportions were compacted at different moisture contents and tests such as unconfined compression (UCS), California bearing ratio (CBR) and hydraulic conductivity tests were carried out.

Kalkan (2006) has carried out a study to determine the suitability of using red mud for soil stabilization. They report, that clay samples in compacted state containing red mud and cement–red mud additives have compressive strength higher than other soil samples.

The effect of combination of different industrial by-products or wastes is not found reported in the literature. There is an urgent to explore the effect of using combination of industrial by-products and wastes on the geotechnical strength parameter.

In this research, experimental studies were conducted out to determine the possible use of combination of different industrial by-products or wastes such as foundry sand, fly ash and red mud as highway and village road construction materials. Two types of stabilizing agents, cement and lime are considered in this study. Unconfined compressive strength (UCS) of soil mixtures having different percentages of industrial by-products and wastes were determined at three different curing periods such as 7, 14 and 28 days.

2. MATERIALS AND METHODOLOGY

The materials used in the present experimental study and the methodology adopted was discussed in this section. Waste foundry sand (WFS) was obtained from the nearby foundry industry, fly ash and red mud were obtained from aluminium industry and thermal power plant. WFS approximately had 99% particles that passed 2.36mm sieve. Physical observations indicated that, WFS had no significant cohesive properties. From particle size distribution, the coefficient of curvature and uniformity coefficient were determined to be 1.16 and 2.52, respectively. Thus the WFS used in the presented study is classified as per Uniform Soil Classification system as poorly graded. Effective size of WFS is found to be 0.21mm. Table 1 gives the chemical composite of the WFS.

Table 1. Chemical Composition of WFS

Sl. No.	Compound	%
1	Silica (as SiO_2)	93.54
2	Alumina (as Al_2O_3)	1.89
3	Calcium Oxide (CaO)	0.34
4	Magnesium Oxide (MgO)	0.10
5	Manganese (as MnO)	0.019
6	Iron (as Fe_2O_3)	1.03
7	Titanium (as TiO_2)	0.17
8	Phosphorus (as P_2O_5)	0.018

9	Potassium (as K_2O)	0.064
10	Sodium (as Na_2O)	0.12
11	Loss on ignition	2.61

Class F fly ash was used in this study. Fineness modulus of the fly ash is 0.058 and the specific gravity is 2.3.

The red mud used was obtained from Aluminium industry named M/S Malco located at Salem, Tamilnadu, India. Chemical composition of fly ash and red mud used were studied.

The cement used is Ordinary Portland Cement of 53 Grade conforming to the IS 12269 (1987). The lime used was obtained by heating and grinding seashells. The seashells were obtained from Kanyakumari district, TamilNadu, India. Mixture proportions considered for this experimental investigation are given in Table 2.

Table 2. Mixture Proportions Considered for UCS Tests

ID	Industrial wastes/by-products			Stabilizing material	
	WFS	Fly Ash	Red mud	Cement	Lime
UC1	85	10	0	5	0
UC2	80	15	0	5	0
UC3	75	20	0	5	0
UC4	70	25	0	5	0
UC5	65	30	0	5	0
UC6	60	35	0	5	0
UC7	85	5	5	5	0
UC8	80	10	5	5	0
UC9	75	15	5	5	0
UC10	70	20	5	5	0
UC11	65	25	5	5	0
UC12	60	30	5	5	0
UC13	85	0	10	5	0
UC14	80	10	10	5	0
UC15	75	15	10	5	0
UC16	70	20	10	5	0
UC17	65	25	10	5	0
UC18	60	30	10	5	0
UC19	85	10	0	0	5
UC20	80	15	0	0	5
UC21	75	20	0	0	5
UC22	70	25	0	0	5
UC23	65	30	0	0	5
UC24	60	35	5	0	5
UC25	85	0	5	0	5

UC26	80	10	5	0	5
UC27	75	15	5	0	5
UC28	70	20	5	0	5
UC29	65	25	10	0	5
UC30	60	30	10	0	5
UC31	85	0	10	0	5
UC32	80	5	10	0	5
UC33	75	10	10	0	5
UC34	70	15	10	0	5
UC35	65	20	10	0	5
UC36	60	25	10	0	5

Optimum Moisture Content (OMC) and the corresponding dry unit weight of the mix proportions were determined using the guidelines of ASTMD1557-12e1. Unconfined compressive strength (UCS) is determined by using the guidelines of ASTM D1633-07. The mix proportions of industrial wastes and by-products considered are compacted at OMC. The compacted samples are wrapped and cured for 7-day, 14-days and 28-days.

3. RESULTS AND DISCUSSION

The results of the experiments are given in Table 3.

Table 3. UCS Test Results of Mixtures Considered

ID	UCC (kN/m^2)		
	7 days	**14 days**	**28 days**
UC1	107	302	427
UC2	213	444	551
UC3	320	658	729
UC4	462	784	889
UC5	384	640	711
UC6	320	534	604
UC7	211	409	640
UC8	320	617	818
UC9	489	818	1069
UC10	622	960	1134
UC11	498	782	960
UC12	462	676	853
UC13	178	498	711
UC14	284	711	889
UC15	424	960	1134
UC16	551	1067	1244
UC17	452	853	1089
UC18	419	747	995
UC19	76	107	121

UC20	107	150	176
UC21	143	211	234
UC22	171	231	256
UC23	156	220	231
UC24	124	178	211
UC25	171	263	302
UC26	185	370	420
UC27	231	462	533
UC28	284	533	617
UC29	250	444	484
UC30	220	356	409
UC31	150	224	256
UC32	167	320	370
UC33	196	424	462
UC34	249	462	498
UC35	231	391	424
UC36	185	302	320

3.1 Cement as Stabilizing Agent

With cement as stabilizing agent, effect of use of different percentages of red mud, fly ash and different curing periods on the strength variation is discussed in this section. Fig. 1 shows strengthvariation with different percentages of red mud.

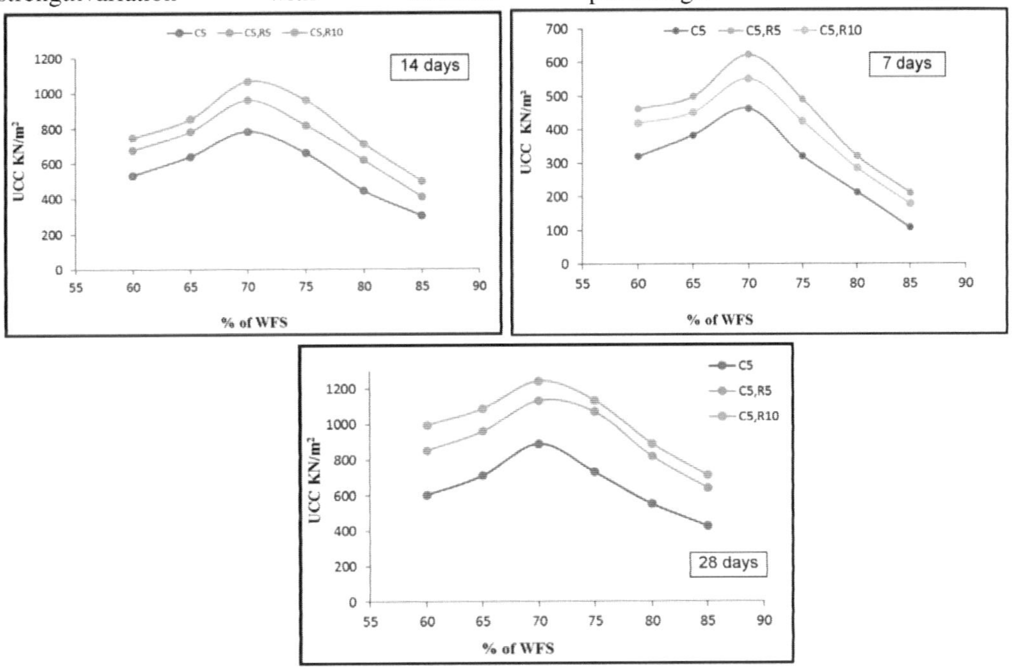

Figure 1. Strength Variation with Red Mud Percentage

From the Fig. 1, it is observed that, the strength increases with increase in the WFS percentage up to 70% beyond which the strength reduces. This may be attributed to reduction in the finer particles in the mixtures that actually form the matrix to hold the relatively coarser particles of WFS. This observation is in general seen for all three curing periods considered. Comparing the curves at different curing period indicates that rate of decrease in the strength is relatively lower for the mixtures which is cured for 28-days than the specimens cured for 7 and 14-days. The effect of the different curing periods in strength may be inferred from Fig. 2.

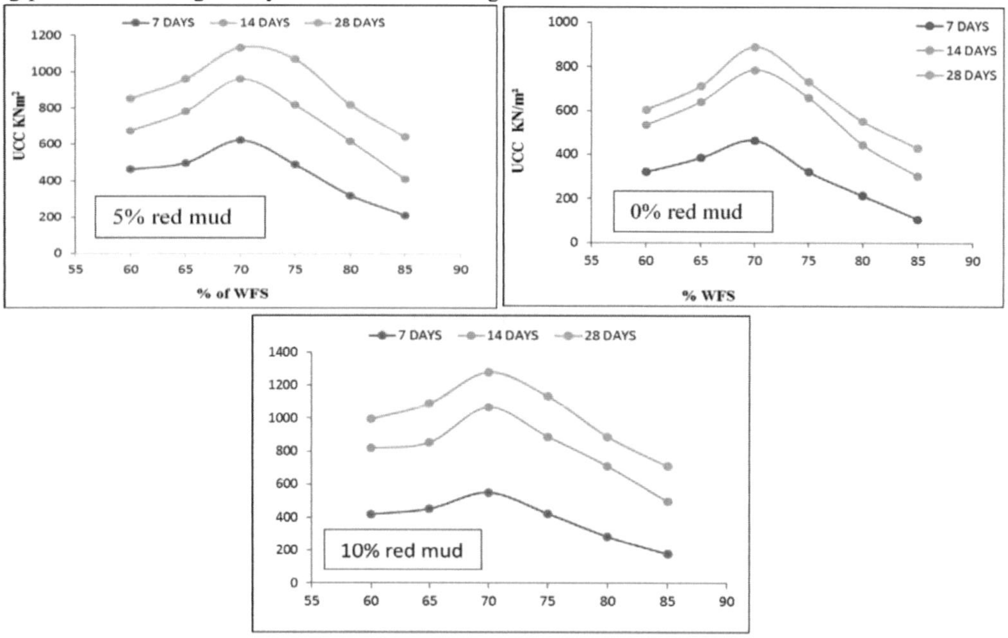

Figure 2. Strength Variation with Curing Period

Fig. 2 it is seen that in general strength increases when curing period is increased. Red mud corresponding to 0%, 5% and 10% with different percentage of fly ash were studied. Fig. 3 indicates that use of 5% red mud corresponds to different amount of fly ash in the mixture.

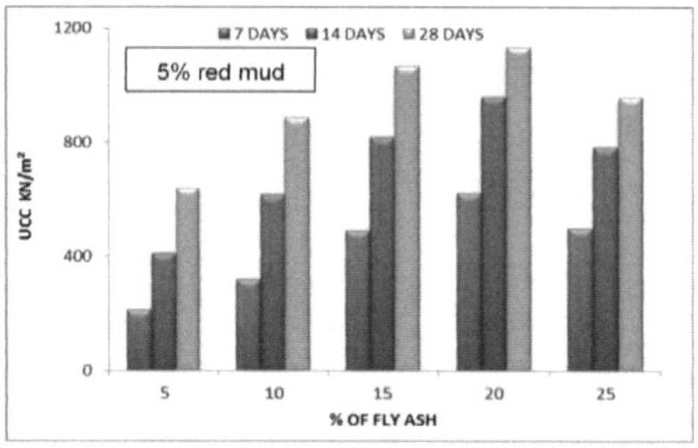

Figure 3. Effect of Fly Ash Percentage on Strength

More than 20 per cent use of fly ash was found to reduce the strength of the mixture. Hence this percentage was considered as optimum.

3.2 Lime as Stabilizing Agent

With lime as stabilizing agent, effect of use of different percentages of red mud, fly ash and different curing periods on the strength variation is discussed in this section. Fig. 4 shows the strength variation with different percentages of red mud used.

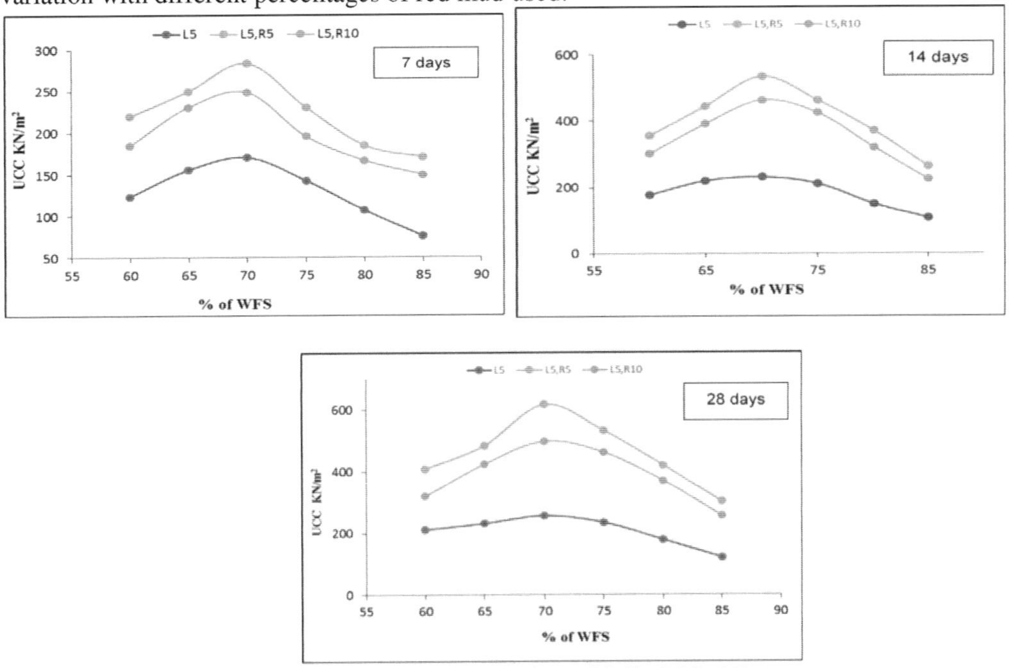

Figure 4. Strength Variation with Red Mud Percentage

Fig. 4 indicates that indicates that use of 5% red mud resulted in higher strength compared to 10% use of red mud. Nevertheless, with respect to cement as stabilizing agent, 5% use of red mud resulted in higher strength. Thus, there seems to be chemical reactions between red mud and cement compounds or compounds formed during hydration. With lime, red mud is less effective under higher per cent of its usage. But with cement, use of higher per cent of red mud increased the strength. From fig.5 the effect of curing periods over the strength can be inferred.

Fig. 5, indicates in general the strength of mixtures increase with the increase in curing period. It is also noted that, the strength corresponding to 14-days of curing is nearly the same as that of 28-days curing.

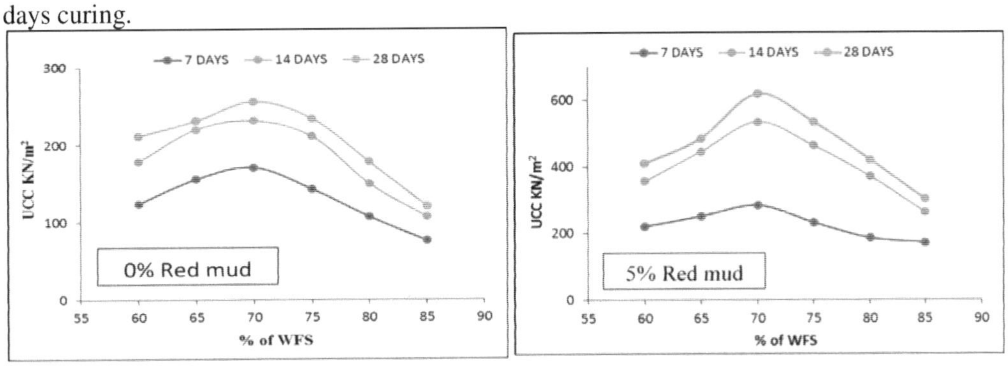

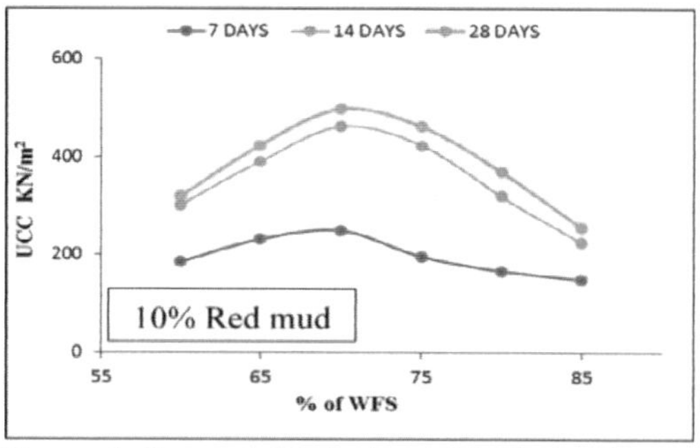

Figure 5. Strength Variation with Curing Period

The effect of use of fly ash in the mixture and the strength at different curing periods was inferred. Based on the result it was found that more than 20% use of fly ash reduces the strength of the mixture same as cement as a stabilising agent.

4. Summary and Conclusions

The experimental study which was carried out to investigate the strength parameter of industrial by-products and wastes such as fly ash, waste foundry sand (WFS) and red mud together with stabilizing agent such as cement and lime were discussed and summarised below.

- Strength of the mixture decreases when WFS percentage was increased beyond 70% and hence 70% of WFS may be considered to be optimum.
- For both stabilizing agents considered in this study, in general it is found that strength of mixtures increases with increase in the curing period.
- When lime is used as stabilizing agent, red mud is found to be less effectual in improving the strength.
- 20% use of fly ash in the mixture may be considered to be an optimum content.
- Use of combined use of industrial by-products and wastes may thus be considered a viable option of disposing them in large quantities. More importantly this technology would ultimately help achieving sustainable development.

REFERENCES

1. Dhawan, PK, Swami, RK, Mehta, HS, Bhatnagar, OP &Murty, AVRS, 1994, Bulk utilization of coal ashes from road works,Indian highways, **22** 11 21-30.
2. Sen, T& Mishra, U 2010 Use of Industrial Waste Products in Village Road Construction *International Journal of Environmental Science and Development* **1** 2.
3. Schroeder, RL 1994 The use of recycled materials in highway construction *Public Roads* **57** 32-41.
4. Sybilski 2004 Use of Industrial Waste Materials in Road Construction in Poland *International RILEM Conference on the Use of Recycled Materials in Building and Structures. RILEM Publications SARL* 351-360.
5. Kirk, PB 1998 Field Demonstration of Highway Embankment Constructed Using Waste Foundry Sand, Ph.D. Dissertation, Purdue University, West Lafayette, IN, 202 p.
6. Mast, DG& Fox, PJ 1998 Geotechnical performance of a highway embankment constructed using waste foundry sand. In: Vipulanandan, C., Elton, D. (Eds.), Recycled Materials in Geotechnical Applications, Geotechnical Special Publication 79. ASCE, Boston, MA, pp. 66-85.
7. Goodhue, M, Edil, TB& Benson, CH 2001 Interaction of foundry sand with geosynthetics *Journal of Geotechnical and Geoenvironmental Engineering* **127** 353-362.
8. Guney, Y,Aydilek, AH &Demirkan, MM 2006 Geoenvironmentalbehavior of foundry sand amended mixtures for highway sub-bases *Waste management* **26** 9 932–945.
9. Prabakar, J, Dendorkar, N&Morchhale RK 2004 Influence of fly ash on strength behavior of typical soils *Construction and Building Materials* **18** 263-267.
10. Kalkan, E 2006 Utilization of red mud as a stabilization material for the preparation of clay liners *Engineering Geology* **87** 220-229.
11. Mazloom, M, Ramezanianpour, AA and Brooks, JJ 2004 Effect of silica fume on mechanical properties of high-strength concrete *Cement and Concrete Composites* **26** 347-357.
12. Meenakshi, SS &Ilangovan, R 2011 Performance of Copper slag and ferrous slag as partial replacement of sand in *Concrete International Journal of Civil and Structural Engineering* **1** 918-927.

13. Nath, P &Sarker, P 2011 Effect of Fly ash on the Durability Properties of High Strength Concrete *Procedia Engineering* **14** 1149-1156.
14. Oner, A, Akyuz, S &Yildiz, R 2005 An experimental study on strength development of concrete containing fly ash and optimum usage of fly ash in concrete *Cement and Concrete Research* **35** 1165-1171.
15. Rahal, K 2007 Mechanical properties of concrete with recycled coarse aggregate Building Environment **42** 407-415.

Batch Adsorption Studies on Treatment of Textile Industry Effluent using Bamboo and Green Coconut shell Activated Carbon

Shantini A. Bokil[1*], Niraj S. Topare[2], Satish V. Khedkar[3]

[1*] School of Civil Engineering, Dr. Vishwanath Karad MIT World Peace University, Pune-411 038, India

[2] School of Chemical Engineering, Dr. Vishwanath Karad MIT World Peace University, Pune-411 038, India

[3] Department of Chemical Engineering, College of Engineering & Technology, Akola-444 104, India

E-mail: shantini.bokil@mitwpu.edu.in

Abstract. The use of activated carbon (AC) as a strong adsorbent was commonly used. This work deals with the study of textile industry wastewater (dye) adsorption on AC produced from Bamboo and Green coconut shell through chemical activation method. The effect of various process parameters such as the amount of adsorbents, initial concentration of dye, pH, contact time, temperature and agitation speed on percentage removal of dye on the activated carbon i.e. Bamboo activated carbon (BAC) and Green Coconut Shell Activated carbon (GCSAC) batch adsorption tests were investigated. Testing was performed for quantitative description of Langmuir and Freundlich equilibrium adsorption isotherms. Based on the experimental findings, the equilibrium was observed in 6 hours. Langmuir isotherm was the best suited for the adsorption of dye, with a maximal adsorption capacity of 142.86 mg / gm. The removal of dye by the use of BAC has been found to be greater than GCSAC.

1. Introduction

Owing to the widespread and growing use of a variety of dyes, contamination from waste water has become a serious environmental threat in recent years. In many industries including textiles, rubber, plastics, printing, leather, cosmetics and others, dyes are commonly used in the manufacture of coloured goods as well. It is estimated that approximately 3% of these colours, which are produced from most of textiles, are dumped into the water system [1]. The ecological structure of the receiving surface water potential is impaired by the waste water, and so the groundwater supplies are definitely disrupted. The majority of dyes used are stable to light and are not biodegradable in the textile industry.

It is important to process it before discharge in the atmosphere to reduce the risk of contamination from this waste. In order to remediate wastewater dyes, various processes have been investigated including photo-catalytic degradation using UV / TiO_2, sound degradation, biologic Fenton treatment, biodegradation, chemical biological integration process, electrochemical process, adsorption process, degradation process with chemical coagulation-flocculation, etc. It has proven to be the most efficacious among the different techniques for its remediation adsorption technique [2-3]. Adsorption is a process of unit activity which refers to the fixation of molecules to the surface. This is because certain solids preferably adsorb another solution on their surfaces from the solution. Especially with different adsorbents, dyes are removed [4].

A variety of these adsorbents were investigated for removal, including the orange peel, the neem seed, the olive cakes, the date palm, the olive shell, the coal etc. In comparison to other adsorbents, however, activated carbon had important removal properties and it is a flexible, high porosity, high surface adsorbent. Selectively the activated charcoal has been used for removing dye, commonly used in the textile field, among different forms of dyes extracted [5]. The purpose of this study is therefore to evaluate the impact of the process parameters such as the amount of adsorbents (0.5, 1, 1.5, 2, 2.5 g/L), initial concentration of dye (200, 400, 600, 800 mg/L), pH (3, 5, 7, 9, 11), contact time (1, 2, 3, 4, 5, 6, 7, 8 Hrs.), Agitation speeds (100, 200, 300 rpm) and temperature (30, 40, 50 ^{0}C) on removal capacity of dye. However, compared with other conventional adsorbents the adsorption activity of BAC for removal of dye, water has been less reported.

2. Experimental

2.1. Preparation of adsorbents:
The materials necessary for the preparation of the AC (bamboo and green coconut shell) were collected in the local region. To extract soil, dried and crushed seving through a 60-mesh sieve, bamboo and green coconut shell have been washed. The chemical activation agent phosphoric acid (H_3PO_4) was used to generate AC. All Chemicals were used analytical (AR) grade and the glassware was washed with distilled water. The carbonization of the materials was done at 200°C for two hours and allowed to cool at room temperature. A 250 g sample of crushed bamboo and green coconut shell was combined with a fixed volume of H_3PO_4. A bamboo/H_3PO_4 and a green coconut shell/H_3PO_4 mixture were used and kept in the stainless steel container. The container was put in a muffling oven at a temperature of 600^{0}C, reached progressively within 3 hours and was held for 45 minutes. BAC and GCSAC were washed after cooling down until pH 7. In an oven at 100 ° C, the samples were then dried overnight to minimize any moisture content. In an airtight bottle the prepared BAC and GCSAC were stored. Figure 1 shows the flow of processing of raw materials to AC.

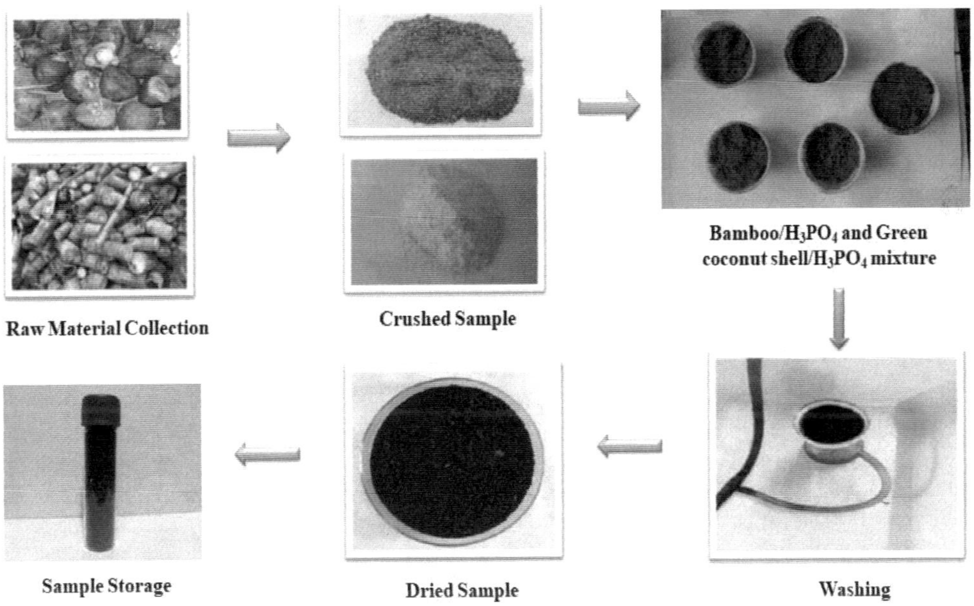

Figure 1. Flow of processing of raw materials to AC

2.2. Characterization of adsorbents:
BAC and GCSAC are the two adsorbents used in the present investigation. Adsorbents used for their characterization, such as bulk density, particle size, surface area and proximate analysis, provide physico-chemical characteristics. Physico-chemical characteristics of the adsorbents are presented in Table-1, BAC bulk density is noted to be lower than GCSAC.

Table 1. Characteristics of the BAC and GCSAC

Sr.No	Properties	Adsorbent Types	
		BAC	**GCSAC**
	Typical Properties		
1	Specific surface area (m^2 g^{-1})	316.20	291.80
2	Total pore volume (cm^3 g^{-1})	0.27	0.31
3	Average pore diameter (A)	34.95	42.72
4	Bulk density (g cm^{-3})	0.60	0.71
	Proximate analysis (wt. %)		
1	Moisture	9.39	10.70
2	Volatile matter	25.57	26.77
3	Fixed carbon	63.37	60.77
4	Ash content	1.67	2.09
	Ultimate analysis		
1	C	61.74	58.74
2	H	1.37	2.24
3	N	5.63	3.07
4	S	9.33	9.99
5	O (by difference)	21.96	21.96

2.3. Experimental Procedure:

2.3.1. Batch adsorption studies:

The tests were performed for the adsorption of the dye using the BAC and GCSAC as adsorbent. Adsorption studies are performed through observation of different important parameters such as namely, amount of adsorbents (0.5, 1, 1.5, 2, 2.5 g/L), initial concentration of dye (200, 400, 600, 800 mg/L), pH (3, 5, 7, 9, 11), contact time (1, 2, 3, 4, 5, 6, 7, 8 Hrs.), Agitation speeds (100, 200, 300 rpm) and temperature (30, 40, 50 ^{0}C). The experimental adsorption studies were conducted using a technique of a completely mixed reactor batch (CMBR). At 200 rpm and 30^0 C, the flasks were agitated in an isothermal shaker to achieve a balance. The time was observed for equilibrium of 5 hours.

The experimental data were therefore calculated up to 8 hours to ensure maximum equilibrium was achieved. For two other flasks containing the same initial dye and the same activated carbon dose were accompanied by identical procedures; however the thermodynamic experiments were studied from 40 to 50^{0}C. The samples were centrifuged for 10 minutes at 1500 rpm in order to avoid interferences from being caused due to carbon fine material and the left-hand concentration in the overnatant solution was analyzed using the UV visible spectrophotometer by monitoring the changes in absorption at a maximum wavelength The dyes were taken from each dyes dye solution at a predetermined time interval [6]. Using the corresponding UV-visible spectrophotometers, dye concentrations in the sample solutions were determined before and after adsorption. It was measured using the normal calibration curve to the dye concentration [7]. In equivalent conditions each test was duplicated and mean values for additional calculations were used. The adsorption value at equilibrium qe (mg / g) was determined accordingly:

$$q_e = \frac{(C_0 - C_e)V}{X} \qquad (1)$$

Where, qe = Adsorbed dye per unit adsorbent mass (mg/g), C_0 = Initial concentration of dye (mg/L), C_e = Final concentration of dye (mg/L), V = Volume of Solution (ml), X = Dose of dry adsorbent used (gm). For the adsorption experiments, fixed amount of adsorbents Bamboo Activated carbon (BAC) and Green coconut shell Activated carbon (GCSAC) was added to 250 mL of dye in each run at definite pH. In a fixed time interval the suspension was adsorbed. A sample was taken using a pipette at various intervals.

3. Result and Discussion

3.1. Influence of operating parameters

The evaluation in the design of all adsorption systems is of vital importance for various operational parameters such as adsorbent dosage, dye, contact time, temperature, pH, agitation speed, etc . This effect was investigated for the removal by BAC and GCSAC adsorbents of various dyes.

3.1.1. Effect of Adsorbent dose

A significant parameter is the adsorbent dose since this defines the adsorbent potential for the initial adsorbate stage. Effect of BAC and GCSAC on dye removal by keeping some further experimental condition consistent at an initial dyes concentration of 200 mg/L was examined in adsorbent doses. Figures 2 and 3 demonstrate the effect of the adsorbent dose on dye removal. The results showed that the removal by percentage has increased to a 2 g/L adsorbent dose and remains virtually constant from Figures 2 and 3. This could result from an improvement in the adsorption dose that increases the adsorption region, which enhances the availability of more sites for adsorption. At low concentrations, equilibrium was found to be reached more easily. Therefore, the adsorbent BAC and GCSAC adsorption dose was chosen 2 g/L for all the adsorption experiments

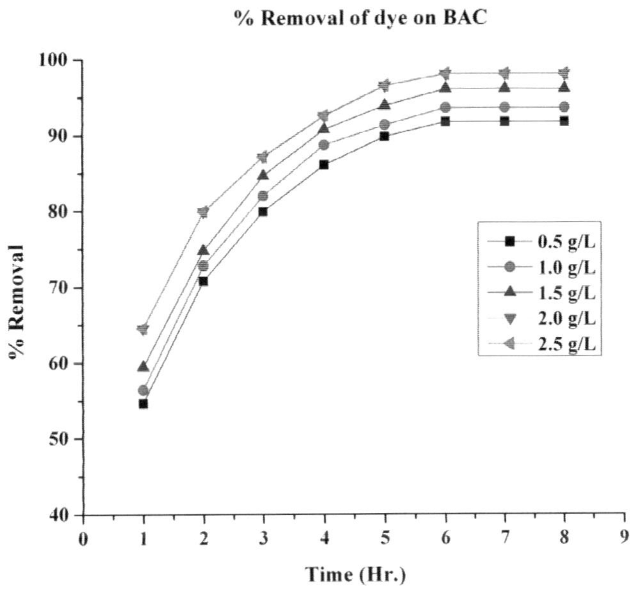

Figure 2. Adsorbent (BAC) dose effect on dye removal

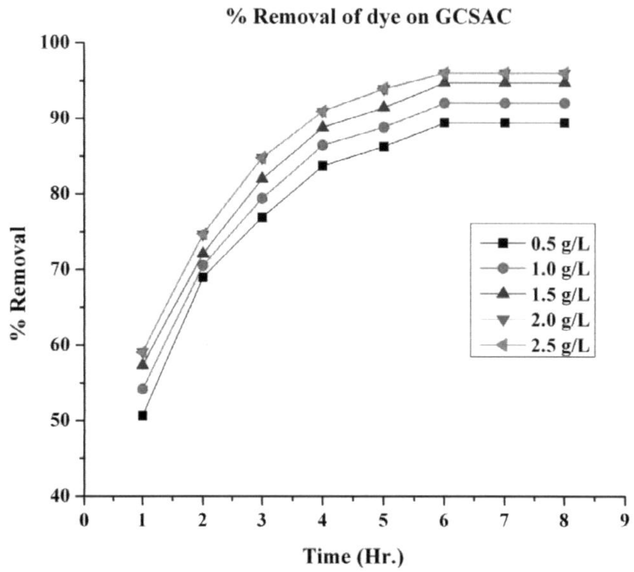

Figure 3. Adsorbent (GCSAC) dose effect on dye removal

3.2.2. Effect of initial concentrations of dye and contact time

At different initial dye levels (200, 400, 600, 800 mg/L) at different time intervals, the experimentation was carried out at 30^0C, with agitation speed at the pH 7 and at a fixed adsorbent dose (2 g/L). The dye removal in various concentrations on BAC and GCSAC is shown in Figure 4 and 5 respectively. The efficiency of the waste water treatment adsorbent is indicated by rapid absorption of contaminants and the establishment of equilibrium in a short time. Even the initial adsorbent or pollutant concentration has an important function since the adsorbent mass will adsorb a fixed amount of solvent only. The further solution or effluent is concentrated, the less effluent amount that can be purified by a specific mass of adsorbent [8]. The fractional adsorption is low in high concentration levels.

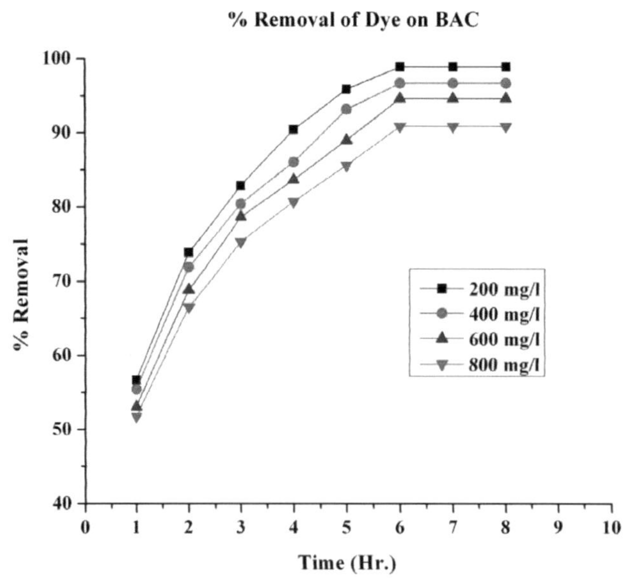

Figure 4. Initial concentration effect and time of contact with dye on BAC for percentage removal

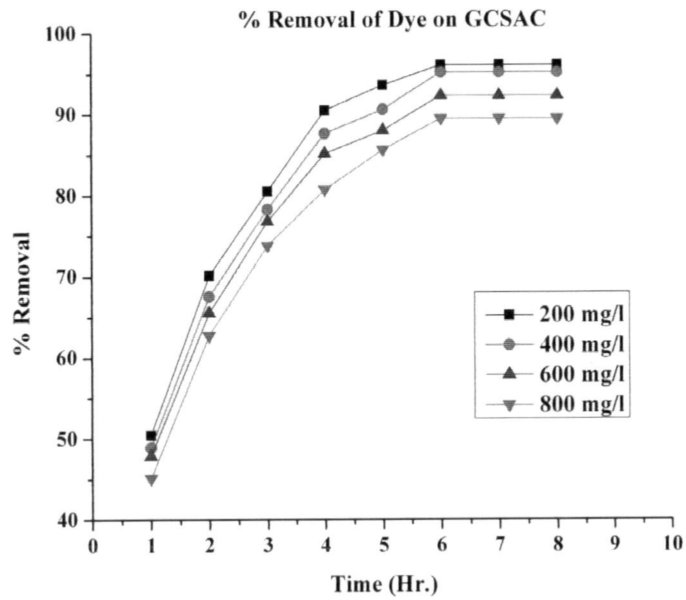

Figure 5. Initial concentration effect and time of contact with dye on GCSAC for percentage removal

It has been observed from Figures 3 and 4 that the removal of the dye is rapid for the first 04 hours and is slower and eventually saturated. A fast adsorption on the outside, followed by slower adsorption on the pores may explain this. Due to the fact that at lower concentrations almost all the colouring molecules were easily assorbed to the outer surface, but further increase in the initial dye concentration, since the initial dye concentration increases from 200 up to 800 mg/L, dye balance was declining from 98.90 to 90.90 %, and dye removal decrease from 96.07 to 89.44 %. The time of equilibrium was found to be independent of the initial concentration of dye. However, it takes 6 hours to reach equilibrium. To ensure complete equilibrium, experimental data were calculated up to 8 hours. The findings show that the removal percentage increases with an initial concentration reduction. However, with an increase in the initial dye concentration in all situations, dye uptake per adsorption unit weight was found to increase.

3.1.3. Effect of pH

Dyes-containing wastewater will be discharged at a different pH; therefore the effect of pH on dye degradation is necessary to analyse. In the whole of the adsorption process and especially in terms of adsorption capability, the pH of the dye solution affects the surface load of the adsorbent, the level of ionisation of the material in the solution and the dissociation of the working groups on the adsorption's active site [9]. It is well known that, due to the presence of H+ ions, the adsorbent is well adsorbed by the anion to lower pH values. Cations are adsorbed at high pH values because of negative adsorption surfaces. Experiments were performed at different pH values ranging from 3, 5, 7, 9 and 11 in order to examine the influence of pH in the removal percentage. The analysis consisted of an agitation by using a water bath shaker at 30^0 C of 250 ml of a dye solution of 50 mg /l with various pH values (3-11). At 200 rpm, 8 hours were disturbed which was more than enough time for equilibrium. The pH was calibrated using the pH metre and with the solutions 0.1 N N NaOH and 0.1 N HCl. The percentage dye removal in pH function is shown in Figures 6 and 7. At higher pH, dye removal is favourable. With pH rising from 3 to 5, the removal of dye is negligible. The removal of the dye increases significantly above pH 5. The maximum removal of dye was observed to be at pH 11. However there is sharp increase in dye uptake for dye from pH 7 to 9 and then after there is little increase in dye uptake as the pH is further increased from 9 to 11. Therefore the best pH range for adsorption of dye was from 7 to 11. There were also findings showing that dye adsorbent effects on BAC and GCSAC of contact time and pH were studied. No effect of pH on equilibrium time, which remained around 6 hrs, was observed.

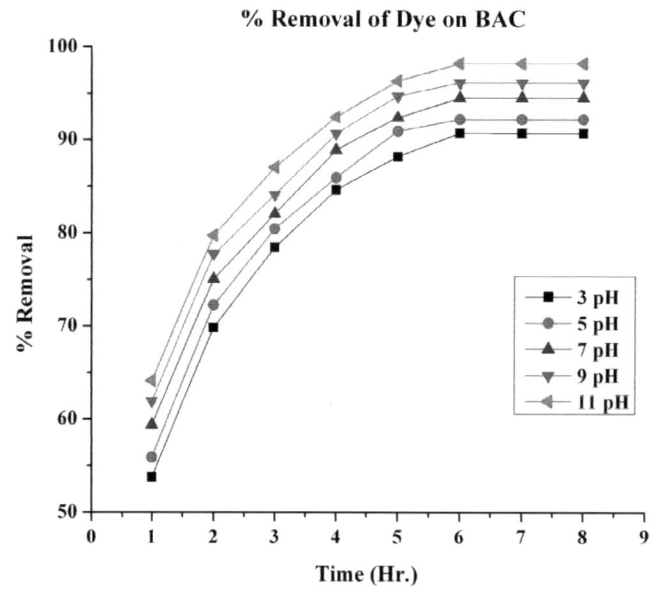

Figure 6. pH effect and contact time on BAC for percentage removal

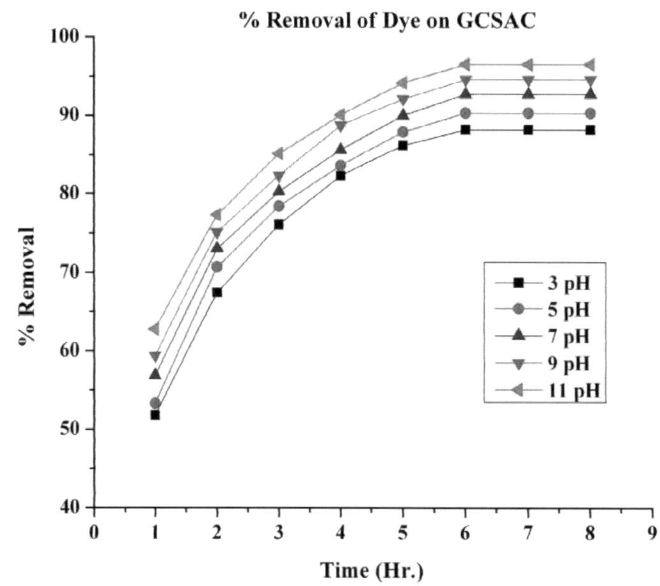

Figure 7. pH effect and contact time on GCSAC for percentage removal

3.1.4. Effect of Agitation speed

Agitation is an useful characteristic for the delivery of solutions in bulk solutions and for the creation of external border films during sorption [10,11]. A broad range of agitation speeds (100, 200, 300 rpm) has been tested in order to ensure that concentration, pH and temperature, along with other parameters, are constant during the process. Figures 8 and 9 display graphically the findings obtained. The equilibrium dye in solid phase was influenced by low turbulent speed from the Figures-8 and 9. These findings suggest that adsorption of dye increases from 89.75 to 97.45% for BAC adsorption, while dyes adsorption of GCSAC increases from 87.24 to 95.95% with rotational speed increases from 100 to 300 rpm. Even so, the absorption of dye is not significantly hampered by increasing the agitation speed from 200 to 300 rpm. In relation to a low speed agitation, however, higher speed equilibrium was achieved.

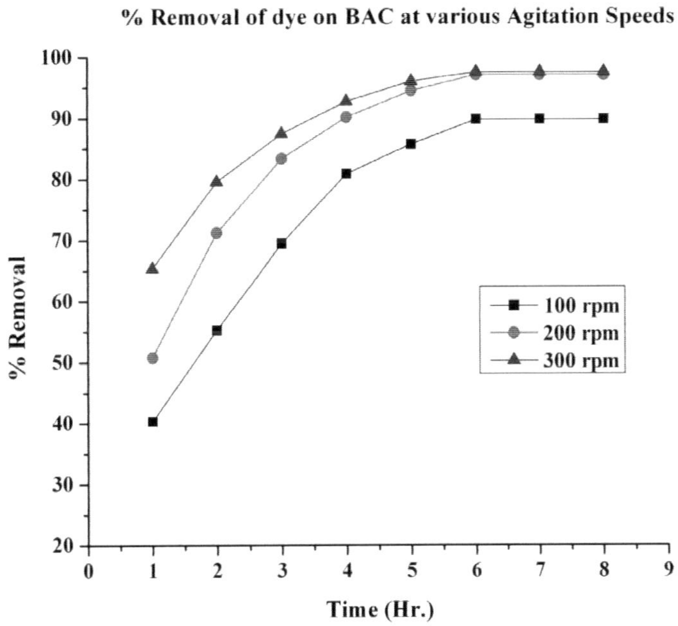

Figure 8. Agitation speed effect and contact time on dye removal on BAC

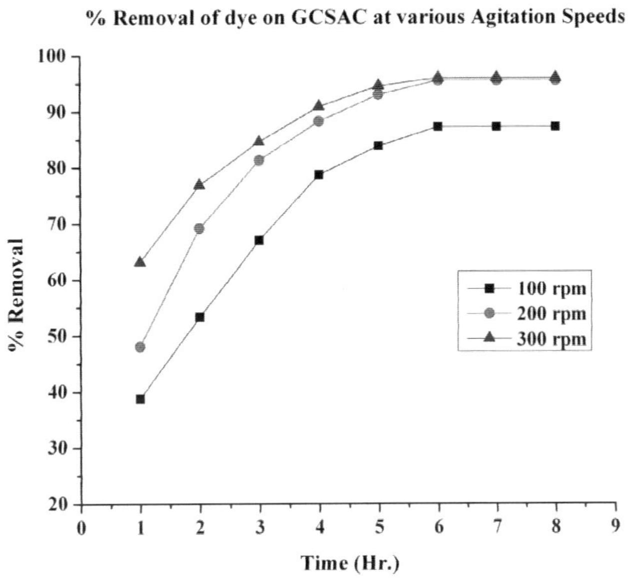

Figure 9. Agitation speed effect and contact time on dye removal on GCSAC

3.1.5. Effect of Temperature

The study of influence of temperature on the rate of adsorption is helpful in the determination of increase or decrease in the extent of adsorption and in computing the corresponding activation energies. Several researchers documented adsorption decreases with temperature increases [12,13]. Adsorption experiments were performed in various temperatures (30^{0}C, 40^{0}C and 50^{0}C) at the initial 200 mg/L dye concentrations at pH 7 and 200 rpm.

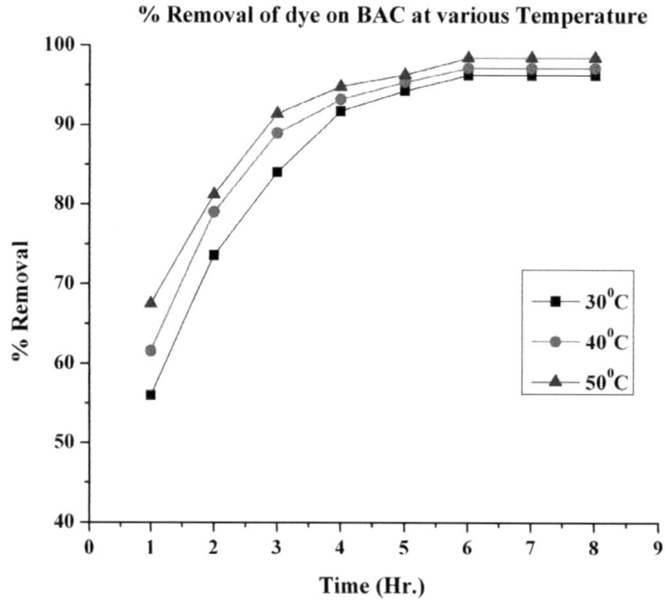

Figure 10. Temperature effect and contact time on BAC dye removal

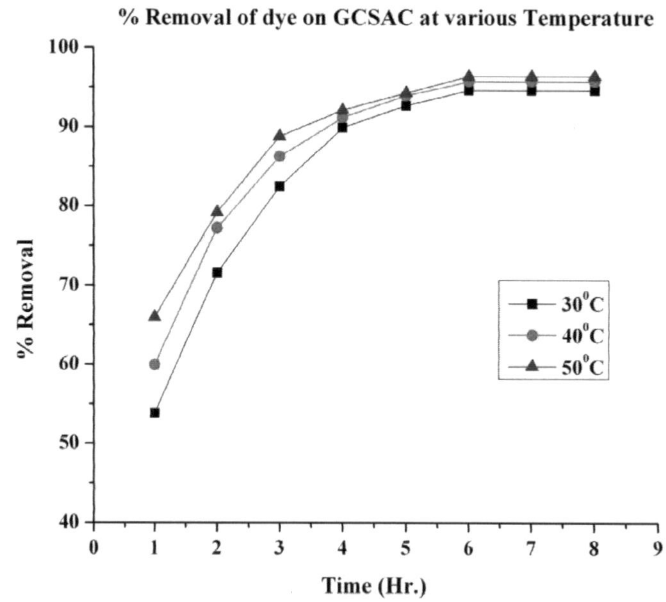

Figure 11. Temperature effect and contact time on GCSAC dye removal

Figures 10 and 11 demonstrate graphically the outcomes of dye absorption experiments with various adsorbents (BAC and GCSAC) at different temperatures (30^{0}C, 40^{0}C and 50^{0}C) and it shows that the BAC and GCSAC adsorbent on removal of dyes is increasing as temperature is rising. The percentage removal of by BAC adsorbent increases from 96.25 to 98.44 % where as it increases from 94.64 to 96.39 % in case of GCSAC as adsorbent with the rise of temperature from 30 to 50^{0}C at 200 mg/l. There is negligible uptake of dye with increase in temperature from 30 to 50^{0}C in view of this all the experiments in batch mode were carried out at 30^{0}C. From these figures, it is apparent that the adsorbed species take up relatively quickly at the initial stages of the adsorption process, which is slow in the later stage and ultimately saturated [14]. Furthermore, these curves are smooth, continuous and saturation-focused.

3.2. Isotherm results:

3.2.1. Langmuir Isotherm results:
Equilibrium data were analysed with Langmuir's model to absorb dye onto BAC and GCSAC. The product with unique adsorption (C_e / q_e) in relation to the equilibrium concentration (C_e) has a linear relationship shown in Figures-12-(a, b , c) and 13-(a, b , c) with correlation factors (R^2) > 0.99 which confirm that the Langmuir model has the same adsorption energy as BAC and GCSAC adsorbents. Form the graph, the slope was found which gives us the value of $1/(q_{max}K_L)$. The equation is y = mx + c. The y-axis intercept gives the $1/q_{max}$= m. Maximum q_m monolayer dye capacity obtained with Langmuir model for dyeing BAC adsorbent is 142.86 mg / gm and 125 mg / gm dye with GCSAC adsorbent at 300^C. Table 2 and 3, represent the Langmuir Models Isotherm constants for adsorption of dye on BAC and GCSAC.

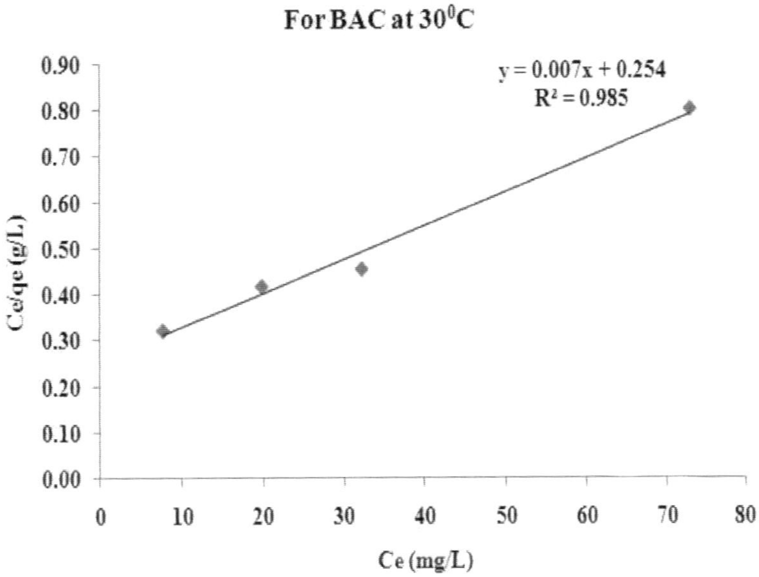

Figure 12 (a). Langmuir Isotherm plot for dye on BAC at 30^{0}C

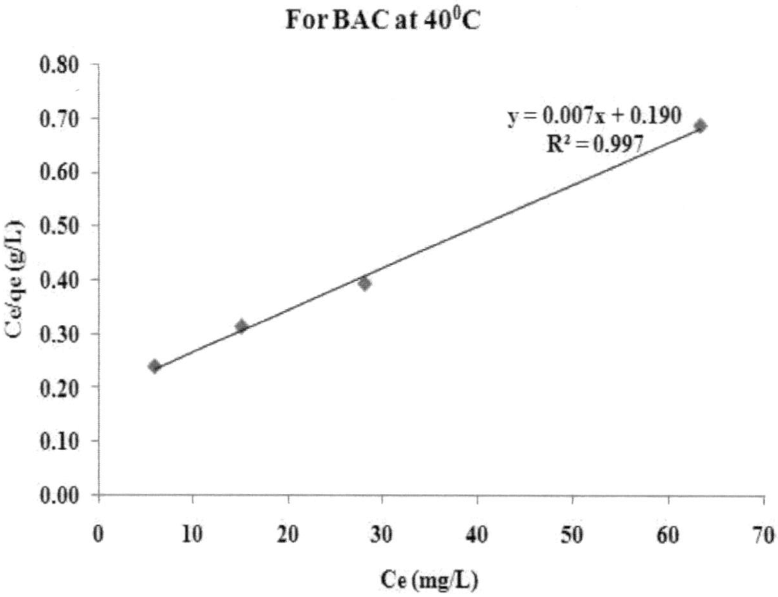

Figure 12 (b). Langmuir Isotherm plot for dye on BAC at 40^0C\

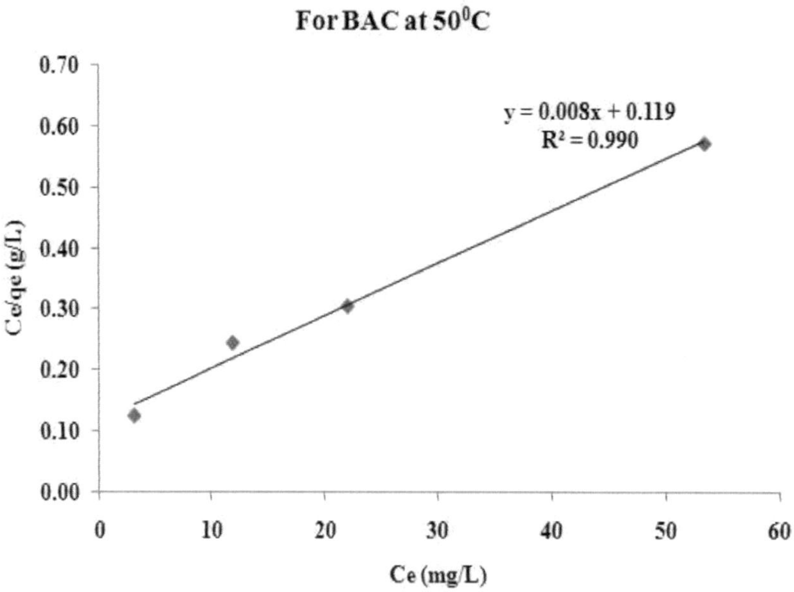

Figure 12 (c). Langmuir Isotherm plot for dye on BAC at 50^0C

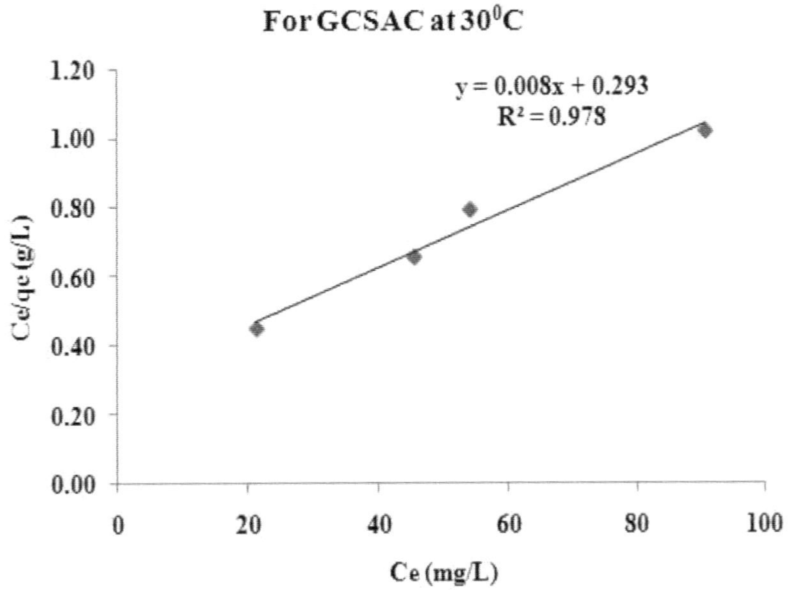

Figure 13 (a). Langmuir Isotherm plot for dye on GCSAC at 30^0C

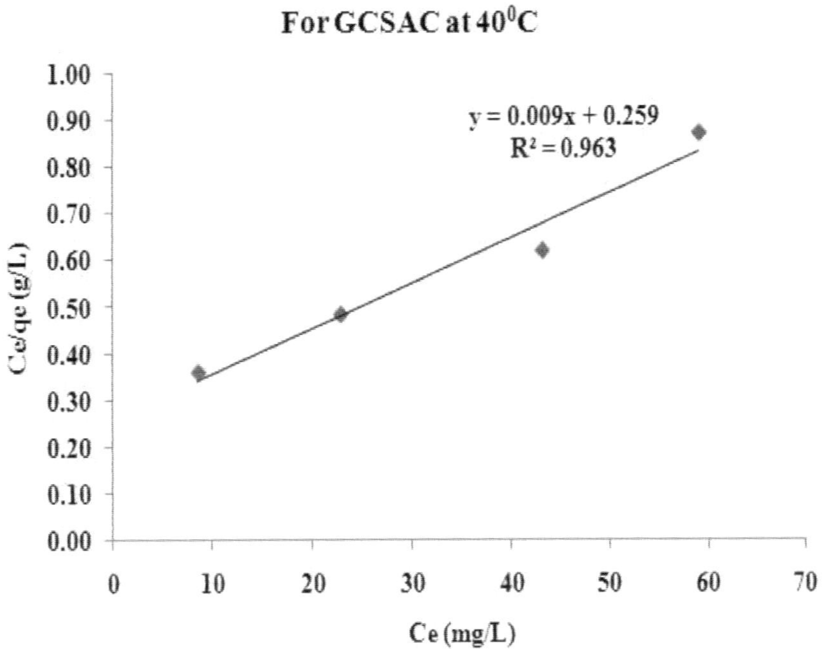

Figure 13 (b). Langmuir Isotherm plot for dye on GCSAC at 40^0C

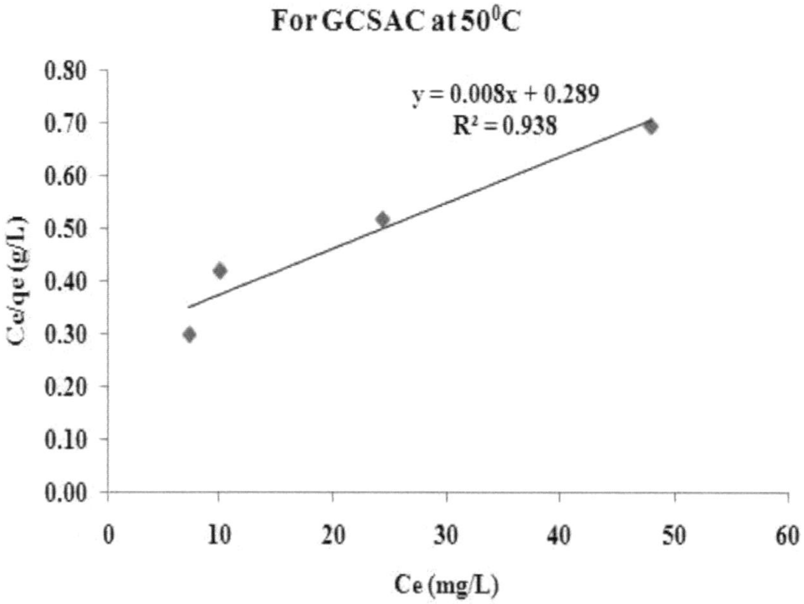

Figure 13 (c). Langmuir Isotherm plot for dye on GCSAC at 50^{0}C

Table 2: Langmuir Models Isotherm constants for adsorption of dye on BAC

Parameters for the Langmuir adsorption isotherm onto BAC		
Temperature (^{0}C)	**Constants**	**Magnitude**
	$K_{L,}$ [L/mg]	0.024
	q_{max} [mg/g]	142.86
	R^2	0.985
	$K_{L,}$ [L/mg]	0.029
	q_{max} [mg/g]	142.86
	R^2	0.997
	$K_{L,}$ [L/mg]	0.055
	q_{max} [mg/g]	125
	R^2	0.99

Table 3. Langmuir Models Isotherm constants for adsorption of dye on GCSAC

Parameters for the Langmuir adsorption isotherm onto GCSAC		
Temperature (^{0}C)	**Constants**	**Magnitude**
	$K_{L,}$ [L/mg]	0.027
	q_{max} [mg/g]	125
	R^2	0.978
	$K_{L,}$ [L/mg]	0.026

q_{max} [mg/g]	111.11
R^2	0.963
K_L [L/mg]	0.026
q_{max} [mg/g]	125
R^2	0.938

3.2.2. Freundlich Isotherm results

The data on equilibrium for the dye adsorption to the BAC and GCSAC have been evaluated with the Freundlich equation. Figures-14-(a,b,c) and 15-(a,b,c) represents the plot of ln q_e versus ln C_e, This line is straight, with an interception value of ln K_F and 1 n slop, presented in Tables 4 and 5. R^2 obtained from the Freundlich model shows that the experimental data can fit into the Freundlich model and n >1 suggests a favorable adsorption mechanism for dye on BAC and GCSAC. On the basis of the R^2 value, however, the Freundlich isotherm seems less successful than Langmuir isotherm models.

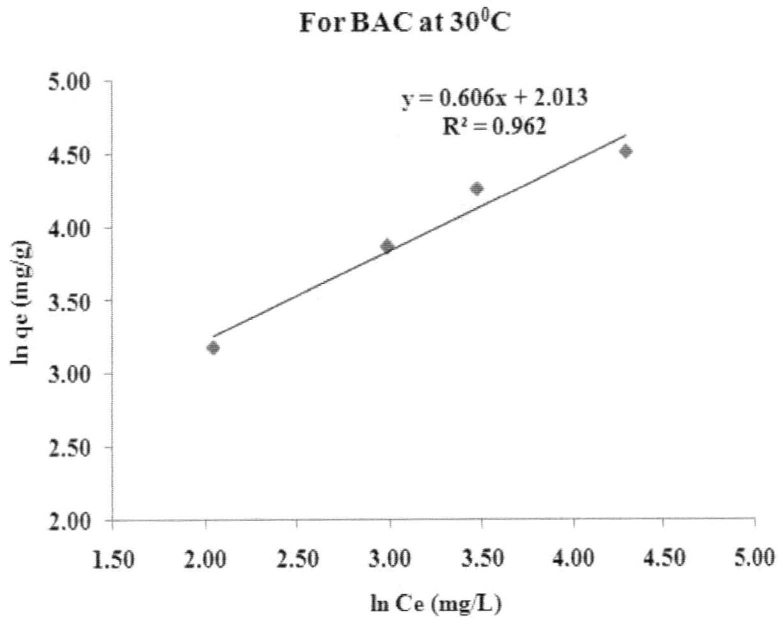

Figure 14 (a). Freundlich Isotherm plot for dye on BAC at 30^{0}C

IOP Conf. Series: Materials Science and Engineering **983** (2020) 012005 doi:10.1088/1757-899X/983/1/012005

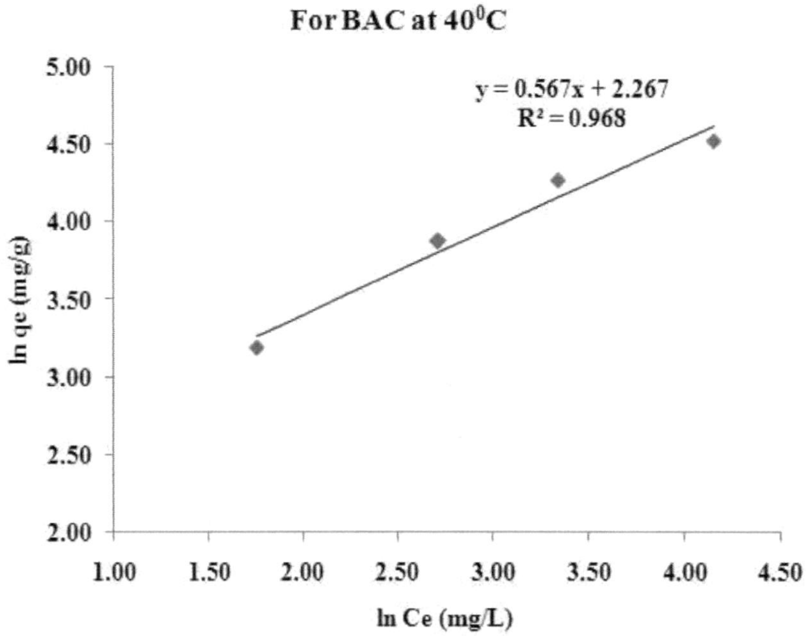

Figure 14 (b). Freundlich Isotherm plot for dye on BAC at 40^0C

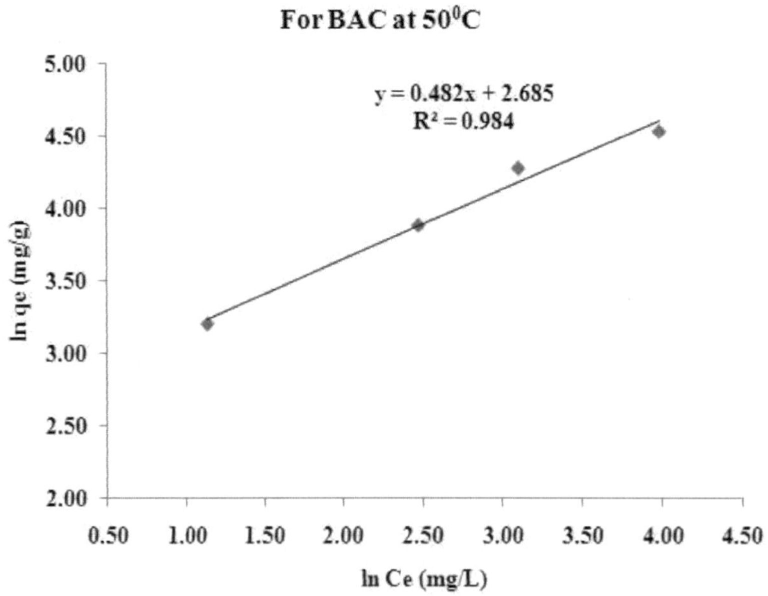

Figure 14 (c). Freundlich Isotherm plot for dye on BAC at 50^0C

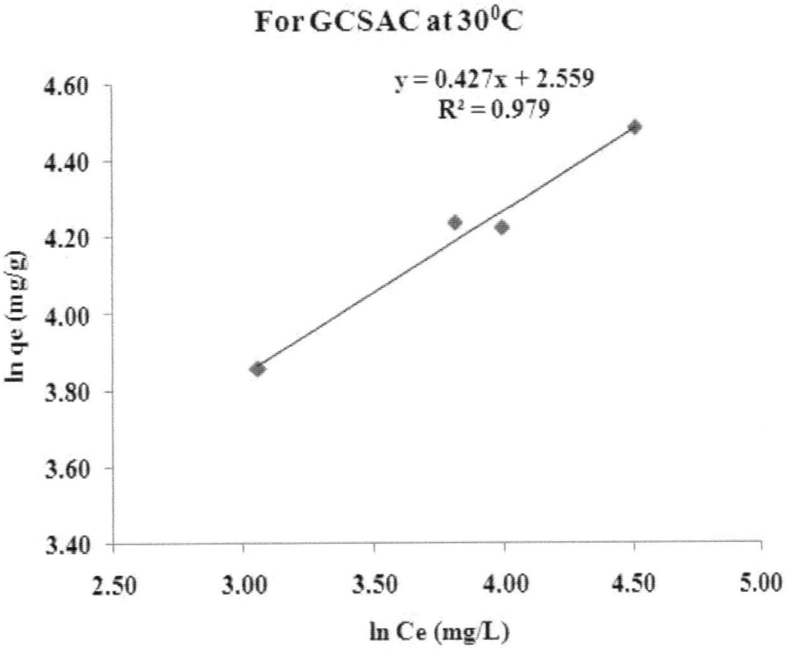

Figure 15 (a). Freundlich Isotherm plot for dye on GCSAC at 30^0C

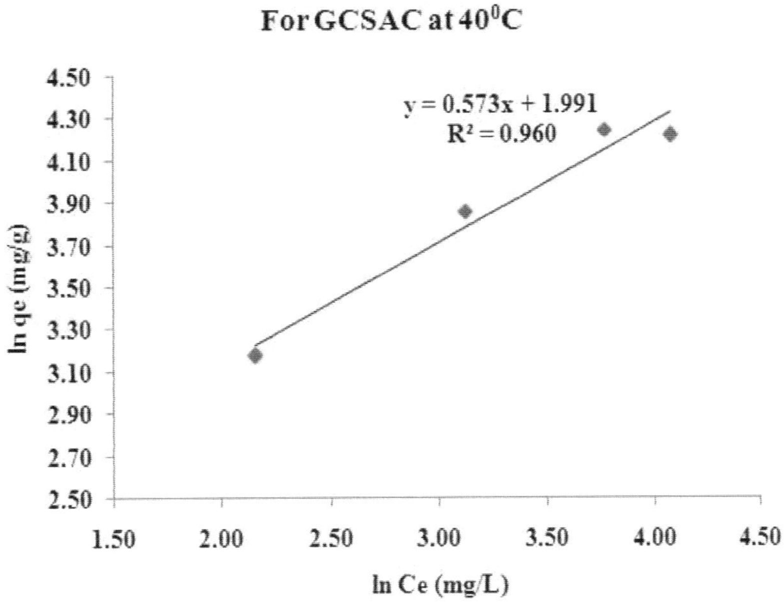

Figure 15 (b). Freundlich Isotherm plot for dye on GCSAC at 40^0C

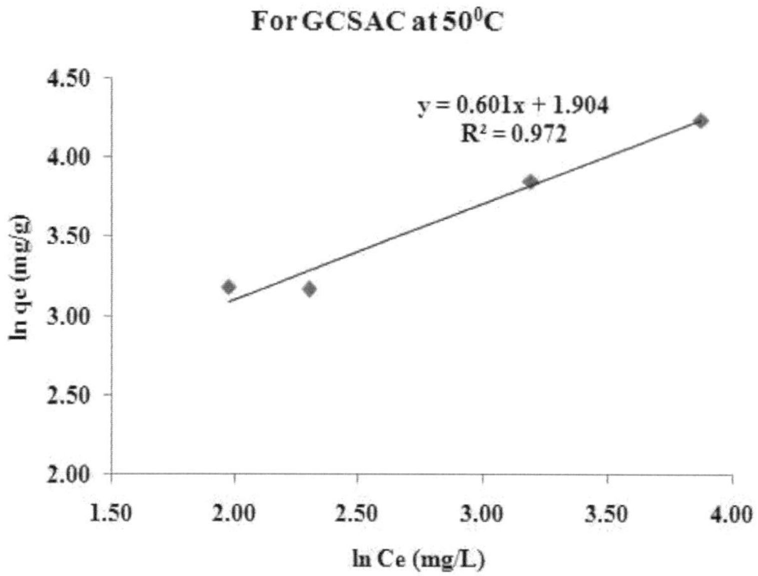

Figure 15 (c). Freundlich Isotherm plot for dye on GCSAC at 50^0C

Table 4. Freundlich Model Isotherm constants for adsorption of dye on BAC

Parameters for the Freundlich adsorption isotherm onto BAC		
Temperature (^{0}C)	Constants	Magnitude
	K_F, [mg g-1] [mg^3 / g]$^{1/n}$	6.76
	n	1.65
	R^2	0.96
	K_F, [mg g-1] [mg^3 / g]$^{1/n}$	8.77
	n	1.76
	R^2	0.97
	K_F, [mg g-1] [mg^3 / g]$^{1/n}$	13.72
	n	2.07
	R^2	0.98

Table 5. Freundlich Model Isotherm constants for adsorption of dye on GCSAC

Parameters for the Freundlich adsorption isotherm onto GCSAC		
Temperature (^{0}C)	Constants	Magnitude
	K_F, [mg g-1] [mg^3 / g]$^{1/n}$	12.96
	n	2.34
	R^2	0.98
	K_F, [mg g-1] [mg^3 / g]$^{1/n}$	6.54

	n	1.75
	R^2	0.96
	K_F, [mg g-1] $[mg^3 / g]^{1/n}$	6.74
	n	1.66
	R^2	0.97

4. Conclusion

The use of BAC as an adsorber is clearly shown to be much more effective than GCSAC in this research work. That can be used effectively to absorb dye from the aqueous solution at 98.90%. During the investigation process, the various parameters observed indicate that the total adsorption process is significantly influenced by the amount of adsorbents, initial concentration of coloration, pH , contact time, agitation speeds and temperature. Test results were adapted for Langmuir as well as Freundlich isotherms. Langmuir is stronger than the Freundlich Isotherm with a maximum adsorption potential of 142.86 mg/gm, demonstrated by adsorption constant and regression factor.

References

[1] Lavarack B.P. 1997 *Hungarian Journal of Industrial Chemistry* **25** 157-160
[2] Topare N.S., Chopade S. G., Khedkar S., Renge V.C,, Bhagat S.L 2011 International *Journal of Advanced Engineering Technology*, **2(3)** 356-358
[3] Tay J. H., Chen X. G., Jeyaseelan S., Graham N. 2001 *Chemosphere*, **44**, 45-51
[4] Topare N.S, Joshi P. 2015 *Emerging Trends in Chemical Engineering*, **2(3)**, 49-51
[5] Srinivasakannan C., Mohamad Z. A. B. 2004 *Biomass and Bioenergy* **27**, 89 – 96
[6] Abd El-Rahim W. M., El-Ardy O. A. M., Mohammad F. H.A. 2009 *Desalination* **249**, 1206-1211.
[7] Alencar, W. S., Acayanka, E., Lima, E. C., Royer, B., De Souza, F. E., Lameira, J., Alves, C. N. 2012 *Chemical Engineering Journal* **209**, 577-588
[8] Al-Husseiny, H. A. 2014 *Journal of Babylon University/Engineering Sciences* **22 (2)**, 296-310
[9] Ali, I., Asim, M., Khan, T. A. 2012 *Journal of Environmental Management* **113**, 170-183
[10] Baseri, J. R., Palanisamy, P. N., Sivakumar, P. *2012 E-Journal of Chemistry* **9(3)**,1122-1134
[11] Ahmad R., Kumar R. 2010 *Applied Surface Science* **257**, 1628-1633
[12] Thakur P., Kumar V. 2019 *J Environ Health Sci Eng*. **17(1)**, 367–376.
[13] Sulyman M., Gierak A. 2020 *Acta Scientific Agriculture* **4(2)**, 1-10.
[14] Ashtaputrey S. D., Ashtaputrey P D. 2016 *Journal of Advanced Chemical Sciences* **2(3)**, 360-362

Comparison of bond strength analysis on the interfacial layer of old and new concrete using latex, epoxy and glycoluril

SKM. Pothinathan[1], M. Muthukannan[2], Narayanan Selvapalam[3],

[1] Department of Civil Engineering, Kalasalingam Academy of Research and Education, Krishnankoil-626126, Tamilnadu, India.

[2] Department of Civil Engineering, Kalasalingam Academy of Research and Education, Krishnankoil-626126, Tamilnadu, India.

[3] Department of Chemistry, Kalasalingam Academy of Research and Education, Krishnankoil-626126, Tamilnadu, India.

Email: s.k.m.pothinathan@klu.ac.in

Abstract. Though other methods were updated for repairing a concrete structure, improving the bond strength of the interfacial layer with effective additives is still the remarkably outstanding technique since it highly improves the structure without any compromise with the quality and strength parameters. The primary reason for the investigation is to compare the market available materials like epoxy and latex with polymer based binding material named glycoluril by slant shear test in both compression and tension. In this comparative study, epoxy bonding depicts higher bond strength in compression than the latex bonding, whereas 3% of glycoluril bonding in the interfacial layer provided for 1cm thick in the interfacial layer were noted to improve the tensile bond strength between the old and new concrete layers which turns to be effective in the repair works of bottom of the slab/beam where tension is more.

1. Introduction

Being the highly used construction material, mortar and concrete obtained from Portland cement are lagging behind because of the lower chemical resistance, limited tensile strength and slower hardening process which induced the development of polymer modified concrete with polymer additives. The long-term durability and higher strength parameters [1] out do the other types of concrete when related to the polymer modified concrete [2].

Polymerization technique can be imparted for the rehabilitation of existing structures by adding up fresh concrete on the surface of the old concrete. Since the old concrete will not have adhesion property and natural bonding, it is overthrown by the usage of bonding adhesives[3,4] Though some techniques are adopted to evaluate the bond strength of the repaired concrete, slant shear method was noted to develop exact stress properties[5,6] and consistent results[7] than the others[8-10]. The slant shear specimens were casted as per ASTM C882[11] in which the specimen is cylindrical in shape with two equal halves casted with a 30^0 verticular diagonal surface. Even though few parameters are responsible for the bond strength between the overlay and the substrate, the surface parameter [12] plays the key role in the interfacial surface behavior.

2. Research significance

Rehabilitation of structures has become one of the challenging fields in concrete industry which promotes a sound structure without harming the previously layered concrete layers along with a perfect bonding between the old and new concrete. However, it is must to strengthen the substrate surface with an appropriate bonding agent which allows the adherence of old and new concrete which in turn extends the service life of the structures. The grooved acid etched specimens significantly performed well when compared with other types of surfaces in a study [13]. The bond strength is usually analyzed to define the bonding between the substrate surfaces and overlay concrete. Tensile strength, slant shear and rebar pull out test were conducted to evaluate the in between bonding of the repaired concrete. Few of them were also adopted to examine the adhesion properties of polymer modified mortar and concrete [14].

3. Experimental study

3.1. Materials

OPC with 43 grade fulfilling IS 8112 necessities was utilized in this investigation. Density of the cement is 3.32g/cm^3 with specific area of 3400cm^2/g. Zone II river sand with specific gravity of 2.56 and coarse aggregate having 20mm maximum size with specific gravity of 2.71 were used. Epoxy and latex were purchased in local market and glycoluril was synthesized according to the standard methodology expressed by Ji-Tai Li [16]. IS10262 used for designing the mix and proportion obtained 191.4 : 510 : 512 : 1157 (W:C:FA:CA) for both old and new concrete.

Three different monomers namely epoxy, latex and glycoluril were used for the polymerization process in the interface surface as show in table 1 for the concrete specimens. When compared with the other two monomers (latex and epoxy), there are very limited study with the glycoluril in polymer modified concrete/mortar for the repairing of the old concrete structures.

Table 1. Concrete test matrix.

Name of the Specimen	Glycoluril (gms)	Epoxy (gms)	Latex (gms)
CS	_	_	_
CCG1%	1.3	_	_
CCG2%	2.6	_	_
CCG3%	3.9	_	_
CCG4%	5.2	_	_
CCE	_	100	_
CCL	_	_	10

CS	Control Specimen
CCCP	Cement Concrete with Cement Paste
CCE	Cement Concrete with Epoxy
CCL	Cement Concrete with Latex
CCG 1%	Cement Concrete with Glycoluril of 1%
CCG 2%	Cement Concrete with Glycoluril of 2%
CCG 3%	Cement Concrete with Glycoluril of 3%
CCG 4%	Cement Concrete with Glycoluril of 4%

3.2. Specimen preparation

In order to plot the bond strength of the interfacial layer in tension and in shear, the splitting tensile strength test and slant shear test were conducted respectively. Three sets of cylindrical specimens

(75mm diameter and 150mm height) were casted for each reading to obtain the level of bonding between the old and new concrete. The old concrete specimen was casted as half cylinders and water cured for 7, 14 and 28 days which is then completed with fresh concrete along with monomers (latex, epoxy and glycoluril of 1%, 2%, 3% and 4%) embedded in between. Epoxy and latex are applied on the surface of the old concrete using a brush (Fig.1.) and then the new concrete is poured in continued by compaction on the surface. Whereas glycoluril is mixed along with cement paste and applied for 1cm thick on the surface of the old concrete followed by the completion of the cylindrical specimen with fresh concrete. Thus, formed specimens were dipped in the formaldehyde solution for a period of 24 hours which initiates polymerization of the monomer (glycoluril). Water curing is again carried out for the completed specimens for 7, 14 and 28 days. The geometry of the specimens was as such 30^0 slanting with the vertical shear plane as per the ASTM C882 standards. The grade of the concrete chosen for this study was M30

Figure 1. Applying monomer on the half cylindrical specimen

3.3. Bond strength analysis

The bond strength between the old and the new concrete were analyzed through the results obtained from the tensile strength and compressive strength analysis. An average value corresponding to three tests were considered. The compressive strength, tensile strength and bond strength were obtained from the test results of the prepared samples. Compressive strength and tensile strength were calculated from the respective compressive load and tensile load required for the failure of the composite specimen. Whereas the bond strength is calculated as (Maximum load)/(Area of slant surface) for further analysis[15]. The slant shears of the specimens were tested as shown in Fig.2. The cylinder specimens were tested at 7, 14 and 28 days from the day of casting the complete specimen along with the polymerized monomer which is intended to improve the strength parameters of the concrete composite.

Figure 2. Testing of Slant Shear Bond Strength

3.4. Types of failure

Three different types of failure modes were noted in the specimens after the tests were conducted namely the failure of the old concrete, failure of the new concrete and the failure of the interfacial layer as show in figure 3. The types of failure were analyzed both in tension and compression as shown in table 2 and 3.

(a) (b) (c)

Figure 3. *(a) Failure of old concrete (b) Failure of new concrete (c) Failure along the interfacial zone*

Table 2. Failure Modes in Compression

Type of mix	7days	14 days	28days
CCCP	Interfacial	Interfacial	Interfacial
CCE	New Concrete	New Concrete	New Concrete
CCL	Interfacial	Interfacial	Interfacial
CCG 1%	Interfacial	Interfacial	Interfacial
CCG 2%	Interfacial	Interfacial	Old Concrete
CCG 3%	New Concrete	New Concrete	New Concrete
CCG 4%	New Concrete	New Concrete	New Concrete

Table 3. Failure Modes in Tension

Type of mix	7days	14 days	28days
CCCP	Interfacial	Interfacial	Interfacial
CCE	Interfacial	Interfacial	Interfacial
CCL	Interfacial	Interfacial	Interfacial
CCG 1%	Interfacial	Interfacial	Interfacial
CCG 2%	Interfacial	Interfacial	Interfacial
CCG 3%	New Concrete	New Concrete	New Concrete
CCG 4%	New Concrete	New Concrete	New Concrete

The differential shrinkage of concrete in interfacial layer due to the different ages of old and new concrete resulting increase in stress. Due to the development of stress at interface most of the repaired concrete were shows decreased bond strength. Epoxy and glycoluril (3% & 4%) were noted to have failed with the new concrete which depicts their better resistance in compression whereas glycoluril (3%&4%) specimens alone were noted to have failed with the new concrete showcasing a strong resistance towards the tensile failure.

4. Test results and discussion:

4.1. Bond strength

All the test specimens were tested after 7days, 14days and 28days of casting the new concrete. Bond strength was calculated by dividing maximum load and interfacial area. The result shows that the bond strength was increasing by adding binding material in the interfacial surface. Bond strength were notably higher for the specimens developed with an interfacial layer of glycoluril (3%&4%) and epoxy compared with the specimens developed with an interfacial layer of latex and glycoluril(1%&2%) showing the effective bonding of the interfacial layer with epoxy and glycoluril(3%&4%) monomers. The bond strength of the above specimens was also noted to increase with the age of the concrete as shown in the Fig.4. There is no bond strength analysis for control specimen. Just to compare the reduction in compressive strength of CS with the bond strength of repaired concrete this was included in Fig.4.

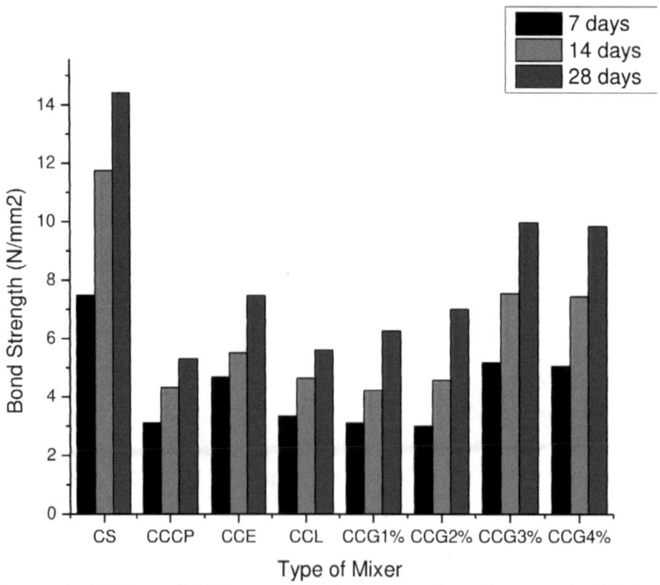

Figure 4. Effect of different specimens on bond strength *(N/mm^2)*

4.2. Compressive strength

The concrete specimens developed with an interfacial layer of the monomer epoxy as well as 3% of glycoluril developed higher compressive strength than the control specimen (M30 mix) at 28 days and 7 days which shows an rise in the compressive strength at early stage and later stage owing to the better hydrogen bonding thus reducing the porous nature of the respective samples (Fig.5.). Whereas modified specimens with latex in interfacial layer tend to develop lower compressive strength than the control specimen due to poor bonding in the interfacial layer. The latex in concrete affects degree of hydration of concrete. Due to this the strength development of the new concrete was reduced and its result in lower compressive strength. This also confirmed by Ma [17].

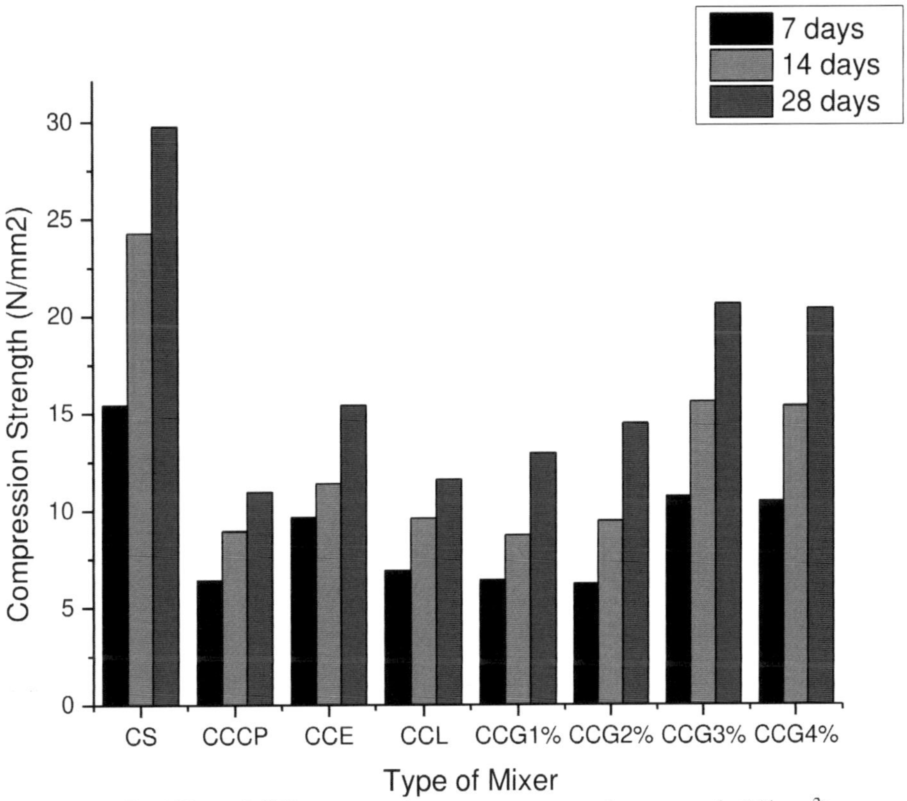

Figure 5. Effect of different specimens on compressive strength (N/mm^2)

4.3. Tensile strength

The tensile strength of all the specimens developed with different monomers as interfacial layers were noted to show a rise along with the age of the concrete (Fig.6.). Notably glycoluril of 3% were noted to develop tensile strength slightly higher than the control specimen whereas specimens developed with epoxy and latex in the interfacial layers resulted in lower tensile strength than the control specimen which depicts the poor bonding of the interfacial layer in the tension zone. This test clearly shows that the glycoluril as binding material plays vital role in tension carrying capacity. Giving equal and slightly hight strength compared to the control specimen. So, this techniques with 3% glycoluril may be advisable for repairing concrete in tension zone like spalling of slab and beam.

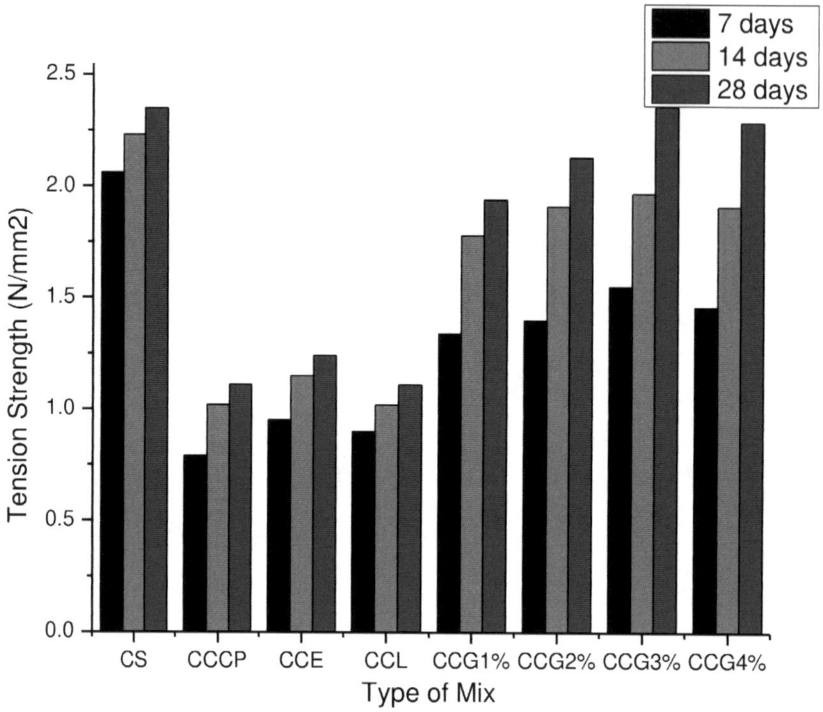

Figure 6. Effect of different specimens on split tensile strength (N/mm^2)

5. Conclusion

Based on the results from the experimental program, the following conclusions were obtained:

- Though concrete and mortar are weak in tension, significant rise in tensile strength were noted in polymer modified concrete with 3% of glycoluril in the interfacial layer which in turn makes it feasible to be of used in the repair works of bottom of the slab/beam where tension is more
- The development of stress due to the differential shrinkage at layer of the repaired concrete shows decreased bond strength and failure at the interface.
- Epoxy and latex developed very less resistance in tension owing to the brittle nature of the respective specimens
- 3% of glycoluril also developed considerable rise in the compression strength along with a noticeable increase in bond strength since the particles are closely packed imparting an efficient hydrogen bonding.
- Repaired concrete with 3% and 4% glycoluril additives in tension strength test giving equal or slightly higher to the control specimen. So, this repair work is advisable in repairing spalling of concrete in slab or beam.

References

[1] Colville, James, Amde M. Amde, and M. Miltenberger 1999 Tensile bond strength of polymer modified mortar. *Journal of materials in civil engineering* **11**, no. 1: 1-5.

[2] Kardon, Joshua B 1997 Polymer-modified concrete. *Journal of Materials in Civil Engineering* **9**, no. 2: 85-92.

[3] Li, Gengying, Huicai Xie, and Guangjing Xiong 2001 Transition zone studies of new-to-old concrete with different binders. *Cement and Concrete Composites* **23**, no. 4-5: 381-387.

[4] Wall, J. S., N. G. Shrive, and B. R. Gamble 1986 Testing of bond between fresh and hardened concrete. In *Adhesion between polymers and concrete/Adhésion entre polymères et béton*, pp. 335-344. Springer, Boston, MA.

[5] Santos, Pedro Miguel Duarte, and Eduardo Nuno Brito Santos Júlio 2011 Factors affecting bond between new and old concrete. *ACI Materials Journal* **108**, no. 4: 449.

[6] Franke, L 1986 The dimensioning of adhesive-bonded joints in concrete building components. In *Adhesion between polymers and concrete Adhésion entre polymères et béton* 461-473. Springer, Boston, MA.

[7] Knab, Lawrence I., and Curtis B. Spring 1989 Evaluation of test methods for measuring the bond strength of Portland cement based repair materials to concrete. *Cement, concrete and aggregates* **11**, no. 1: 3-14.

[8] Elkurdi, Adel, Shafik Khoury, and A. Ali Abdo 2003 Evaluation of new old concrete bonding agent systems by slant shear test In *AICSGE 5*. Faculty of engineering, Alexandria University.

[9] Momayez, A., M. R. Ehsani, A. A. Ramezanianpour, and H. Rajaie 2005 Comparison of methods for evaluating bond strength between concrete substrate and repair materials *Cement and concrete research* **35**, no. 4: 748-757.

[10] Saldanha, R. F. S. M., E. N. B. S. Julio, D. D. Costa, and Pedro MD Santos Modified Slant Shear Test To Enforce Adhesive Failure

[11] ASTM C882 / C882M-13a, Standard Test Method for Bond Strength of Epoxy-Resin Systems Used With Concrete By Slant Shear, ASTM International, West Conshohocken, PA, 2013, www.astm.org

[12] He, Yan, Xiong Zhang, R. D. Hooton, and Xiaowei Zhang 2017 Effects of interface roughness and interface adhesion on new-to-old concrete bonding. *Construction and Building Materials* **151** 582-590.

[13] Mirmoghtadaei, Reza, Masih Mohammadi, Nasibe Ashraf Samani, and Seyedreza Mousavi 2015 The impact of surface preparation on the bond strength of repaired concrete by metakaolin containing concrete. *Construction and Building Materials* **80** 76-83.

[14] Ohama, Y., K. Demura, H. Nagao, and T. Ogi 1986 Adhesion of polymer-modified mortars to ordinary cement mortar by different test methods. In *Adhesion between polymers and concrete/Adhésion entre polymères et béton* 719-729. Springer, Boston, MA.

[15] Pattnaik, Rashmi Ranjan, and Prasada Rao Rangaraju 2007 Analysis of compatibility between repair material and substrate concrete using simple beam with third point loading. *Journal of Materials in Civil Engineering* **19**, no. 12: 1060-1069.

[16] Ji-Tai Li, Xiao-Ru Liu, Ming-Xuan Sun. 2010 Synthesis of glycoluril catalyzed by potassium hydroxide under ultrasound irradiation Ultrasonics Sonochemistry **17** 55-57.

[17] Hongyan Ma, Zongjin Li 2013 Microstructures and mechanical properties of polymer modified mortars under distinct mechanisms, Construction and Building Materials **47** 579–587.

IOP Conf. Series: Materials Science and Engineering 983 (2020) 012007 doi:10.1088/1757-899X/983/1/012007

Concrete-filled steel tube columns of different cross-sectional shapes under axial compression: A review

Ilanthalir A[1,4], Jerlin Regin J[2], Maheswaran J[3]

[1] Research scholar, Anna University, Chennai, Tamil Nadu, India
[2]Associate Professor, Civil Engineering, St. Xavier's Catholic College of Engineering, Chunkankadai, Tamil Nadu, India
[3]Professor, Civil Engineering, St. Xavier's Catholic College of Engineering, Chunkankadai, Tamil Nadu, India
[4]corresponding author email: tenderleaf26@gmail.com

Abstract. This paper presents a review on the axial compression behavior of different cross-sectional shaped concrete-filled steel tube (CFST) columns such as circular, rectangular, square, elliptical, hexagonal, octagonal and round-ended. A discussion on some special shapes like triangular, D-shape, ¼ circular, fan shape etc. is also done in this paper. Confinement offered by the varying cross-sectional shape steel tubes to concrete core, material strength limits and slenderness limits available in different codes of practices and the proposed yield slenderness limits for rectangular, elliptical and octagonal cross-sections are discussed. The failure modes encountered during the axial compression test on CFST columns are mentioned. An analysis on the parameters that influence the behavior of the columns, design equations in the current codes of practice, modifications made on the code design equations, equations proposed on the determination of compressive strength of different cross-sectional shapes of CFST columns is done. Unified formulas for two different cross-sectional shapes and solid and hollow concrete-filled steel tubes are also included. Conclusions are finally made based on the discussion done.
Key words: CFST, cross-sectional shapes, compressive strength.

1. Introduction

Concrete-filled steel tubes (CFSTs) are being used effectively in the construction industry as columns of mid-rise and high rise buildings, in bridge as piers and piles, submarine pipelines, and other engineering structures. CFST are known for their high ductility, high stiffness, greater energy absorption and, greater seismic and fire resistance [1]. Concrete filled steel tube column comprises a concrete core in hollow steel tube with or devoid of additional reinforcement. A concrete-filled steel tube can have either a solid or hollow concrete core. This paper discusses the literatures on steel tubes having solid concrete core without reinforcement.

Steel tubes of circular, rectangular and square are the regularly employed cross-sections in CFST column construction. The square section is considered as an exceptional type of rectangular section. For architectural reasons, hexagonal, octagonal, elliptical and round-ended shapes are used in CFST. The elliptical hollow section has varying radius of curvature around its perimeter [2]. The hexagonal shape can be regular and dual-axisymmetric. The regular hexagonal shape has equal lengths on all sides with an

IOP Conf. Series: Materials Science and Engineering 983 (2020) 012007 doi:10.1088/1757-899X/983/1/012007

interior angle of 120° whereas; dual-axisymmetric shape has equal lengths on all sides with four interior angles of 135° and two interior angles of 90° in the opposite directions. The regular shape of hexagonal column has higher load carrying capacity than its dual-axisymmetric shape column and its steel tube shape closely resembles a circular tube [3]. Octagonal hollow section lies anywhere between circular and square hollow section and is used as transmission poles, lattice towers and telegraph towers [4]. Round-ended rectangular column (RRCFST) is a hybrid column of two different shapes. It can resist the impact of running water with its smooth rounded corners and hence using RRCFST columns as piers can reduce the force of running water and increase the life time of the bridge using such piers [5]. Columns with special shapes such as L-shape, T-shape, triangular, fan-shape, D-shape, 1/4 circular, semi-circular, round-ended, rhombic, T-shape, plus shaped, pentagram, and fan-shape are some of the recent additions to the tubular family.

2. Interaction between concrete core and steel tube

The concrete-steel interface is a region where most of the stress transfer takes place in a composite column. The combined action of steel and concrete in this region is integral as it contributes to high strength and ductility of the column. Steel provides lateral confinement to the infill concrete in a passive way and reduces shrinkage of concrete. Circular steel sections are found to provide the strongest confinement compared to rectangular/square, hexagonal and special-shaped sections such as triangular, Fan, D-shaped, 1/4 circular and semi-circular sections. The confinement of the square/rectangular steel tube to the core concrete is strongest at the corners than at the center. The confinement in elliptical CFST column lies between that in rectangular and circular tube and is a function of the aspect ratio of the section [6]. The confinement effect is assumed to be between those of circular and rectangular CFSTs in hexagonal CFST column and the confinement effect is more at the corner regions than at the center of the core concrete [7]. The confinement is greater at the corners and at the mid-point of the core concrete in octagonal CFST columns [8] and is found to be superior to square CFST columns [9]. For round-rectangular CFST columns, the concrete area in the rounded ends of the column has the strongest confinement and that in the straight edge has the weakest confinement [10]. Concrete core in special-shaped (triangular, Fan, D-shaped, 1/4 circular and semi-circular sections) CFST columns are also confined by the steel tube and is similar to square /rectangular hollow section columns [11]. The confined areas of concrete infill by steel under axial compression in different cross-sectional shapes are shown in Figure 1.

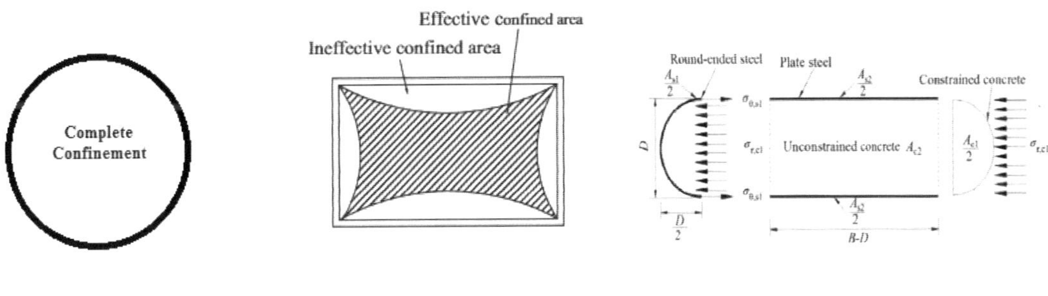

(a) (b) (c)

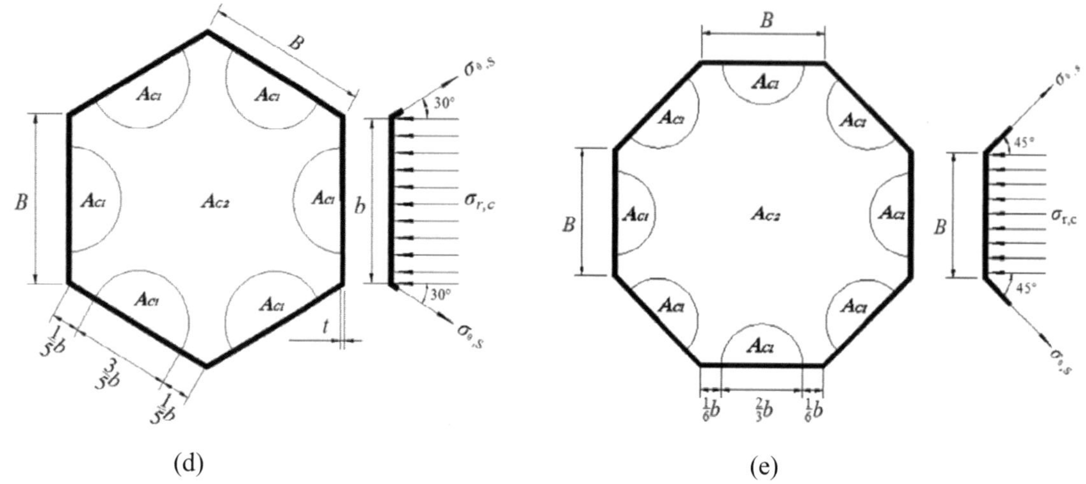

Figure 1. Confined and unconfined areas in different cross-sectional shaped CFST columns under axial compression (a) Circular CFST, (b) Square/Rectangular [19], (c) Round-ended rectangular CFST [10], (d) Hexagonal CFST [13], (e) Octagonal CFST [8].

The arching action of the ineffectively confined concrete in rectangular CFST columns occurs in the form of a second-degree parabola with an initial tangent slope of 45° [12]. The confinement effect in hexagonal steel tube is observed in the corner areas and the center area of core concrete (A_{c2}) while, the length of un-constrained areas (A_{c1}) is approximately 3/5b (b is the edge length of concrete core) [13]. 33% of the total area of concrete core is therefore unconfined [3]. For octagonal CFST columns, the length of unconfined area (A_{c1}) is approximately 2/3b (b is the edge length of concrete core) and the effectively confined area (A_{c2}) in octagonal section is nearly 70% of the total cross-sectional area of the octagonal column [8].

3. Materials for CFST columns

3.1 Hollow steel tubes

Mild steel, cold-formed steel and high strength structural steel are the general types of steel used in CFST columns. The stress-strain curve for mild steel is marked by a typical yield plateau. The stress-strain curve of cold-formed steel is rounded and the rounded stress-strain curve is replaced by a straight line for high strength steel.

Tensile coupon test on steel is generally carried out to determine the material characteristics. Coupons are generally obtained from corner, curved and flatter regions of the hollow tubes depending upon the shape of the tube. Width-to-thickness ratio (B/t) and diameter-to-thickness ratio (D/t) of steel plates are widely used to determine the performance of rectangular and circular CFST columns and as a measure to prevent local buckling. Steel strength is taken either in the form of $\varepsilon = \sqrt{235/f_y}$ or $\beta = \sqrt{E_s/f_y}$. ANSI/AISC 360 [14] divides CFT columns into compact, non-compact and slender sections. Table 1 shows the maximum allowable limits on steel tube yield strength and yield slenderness by few design codes.

Table 1. Yield slenderness limits and yield stress limits in hollow steel sections.

Country	Code	Cross- section Shape	Yield slenderness limits	Yield stress f_y MPa
USA	ANSI/AISC 360 [14] Compact/Non-compact sections	Circular	$\leq 0.15(E_s{}^a/f_y)$	≤ 525
		Rectangular	$\leq 2.26\sqrt{(E_s/f_y}$	
	Non-compact/ slender sections	Circular	$\leq 0.19(E_s/f_y)$	
		Rectangular	$\leq 3\sqrt{(E_s/f_y}$	
Europe	EC4 [15]	Circular	$\leq 90\ (235/f_y)$	230 to 460
		Rectangular	$\leq 42\ \sqrt{(235/f_y}$	
Australia	AS 5100-6 [16]	Circular	$\leq 82\ (250/f_y)$	230 to 400
		Rectangular	$\leq 35\sqrt{(250/f_y}$	
USA	ACI 318 [17]	Circular	$\leq \sqrt{8(E_s/f_y)}$	-
		Rectangular	$\leq \sqrt{3(E_s/f_y)}$	

[a]E_s- Young's modulus of steel

3.1.1 Proposed yield slenderness limits and cross-section classification. Hollow steel sections are generally classified between Classes 1-4. Under axial compression, Class 1-3 sections reach their ultimate loads before local buckling and Class 4 sections encounter local buckling before reaching their ultimate loads. Therefore, Class 1-3 sections or non-slender sections have thicker walls than Class 4 sections or slender sections. It is seen from Table 1 that, limits on slenderness in standard codes of practice are found only for circular and rectangular cross-sections. Therefore, due to the lack of code limits, researchers have proposed slenderness limits for the sections such as elliptical and octagonal sections. An extension of the slenderness limit for rectangular section is also included in this section. Design recommendations in rectangular CFST columns for width-to-thickness (h/t) were proposed by Du et al. [18] for EC4, AISC 360, the Chinese code GB 50936 and, by Du et al. [19] for the Chinese code DB 29-57 to extend the code limits to cover high strength steel. Yield slenderness limits on elliptical hollow sections (EHS) were given by Chan and Gardner [20], Teran and Gardner [21] and Zhao and Packer [22]. A new slenderness limit for octagonal cross-section was given by Zhu et al. [4] and, in accordance with EN 1993-1-1, ANSI/AISC 360-16, ASCE/SEI 48-11 and the Direct Strength Method (DSM) incorporated in North American code AISI 100-16 was given by Chen et al. [23]. Table 2 shows the proposed yield slenderness limits for rectangular, elliptical and octagonal hollow sections.

Table 2. Proposed yield slenderness limits.

Cross-sectional shape	Author	Code	Yield slenderness limits
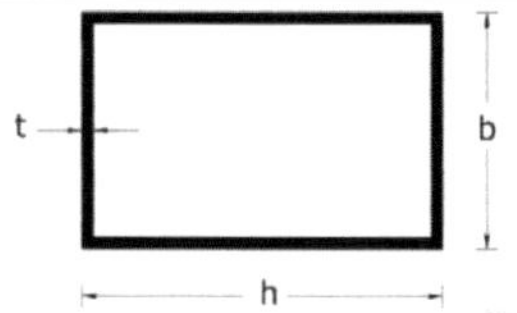 Rectangular	Du et al. [18]	EC4	$h/t\varepsilon = -12\,h/b + 62$
		AISC 360	$h/t\beta = -0.77\,h/b + 2.79$
		GB50936	$h/t\varepsilon = -7.5\,h/b + 65$
	Du et al. [19]	DB 29-57	$\frac{h}{t} = 58.7\sqrt{235/f_y}$
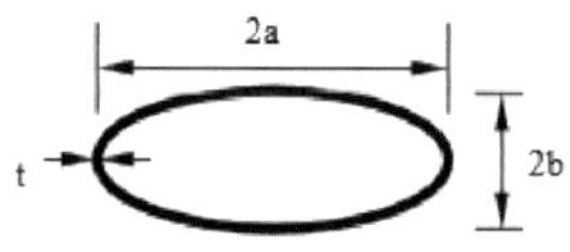 Elliptical [20]	Gardner and Chan [20]	-	$\frac{D_{e1}}{t\varepsilon^2} = 2\frac{(a^2/b)}{t\varepsilon^2}$ Equivalent circular diameter $D_{e1} = 2(a^2/b)$
	Teran and Gardner [21]	-	$\frac{D_{e2}}{t\varepsilon^2} = \frac{2a\left[1 + f.\left(\frac{a}{b} - 1\right)\right]}{t\varepsilon^2}$ Equivalent circular diameter $D_{e2} = 2a\left[1 + f.\left(\frac{a}{b} - 1\right)\right]$ Coefficient $f = 1 - 2.3(t/2a)^{0.6}$
	Zhao and Packer [22]	-	$\frac{D_{e3}}{t\varepsilon} = \frac{2(a-t)}{t\varepsilon}$ Equivalent rectangular hollow section diameter $D_{e3} = 2(a-t)$
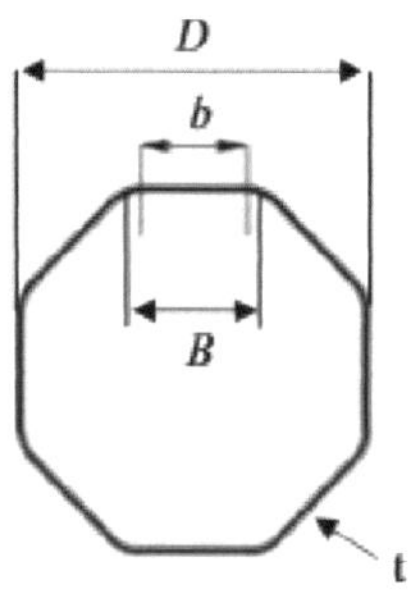 Octagonal [4]	Zhu et al. [4]	-	$b/t \leq 29.8\sqrt{235/f_y}$
	Chen et al. [23]	EN 1993-1-1	$b/t \leq 34.1\sqrt{235/f_y}$
		ANSI/AISC 360-16	$b/t \leq 1.14\sqrt{\frac{E_s}{f_y}}$
		ASCE/SEI 48-11	$b/t \leq 2.01 \text{ x } \frac{260}{\sqrt{f_y}}$
		DSM	$\lambda_p{}^a \leq 0.653$

[a] λ_p – overall cross-section slenderness.

3.2 Concrete

Different strengths of concretes such as normal strength, high strength and ultrahigh strength concrete (UHSC) are generally filled in the steel tubes. Special concretes like self-compacting concrete (SCC) and lightweight concrete (LWC) are also used as infill in steel tubes. SCC in concrete-filled tubes has become popular due to reduced sound and injuries connected to vibration of concrete, higher workability characteristics, higher load capacity, inherent ductility and toughness [7, 22, 24]. LWC has been used in composite construction for its low thermal conductivity and low specific gravity, reduced self-weight of the structure [25]. Lightweight CFST was 15.3% and 26% lighter than normal concrete-filled steel tube columns [1, 25].

The methods to increase the compressive strength of CFST columns are to use large cross-sections and to use high-strength materials. Therefore, the use of high-strength steel tubes with yield strength ≥ 460MPa has attracted much attention. High strength materials in CFST can reduce the size of the member. High strength steel of Q345 or above should be combined with C100 or above grade concrete and a steel ratio not less than 20% is recommended by Zhou et al. [33] for UHSC filled steel tube column to ensure sufficient ductility. For special-shaped CFST columns, the designers are recommended to attempt higher steel tube strengths by Wang and Han [11]. It is suggested to limit the use of expansive agent in ultra-high performance concrete filled steel tubes to 12% by Huang et al. [53]. Table 3 gives the limitations on the concrete compressive strengths in different codes of practice.

Table 3. Compressive strength limits specified in codes of practice.

Code	Concrete strength (MPa) limits
ANSI/AISC 360-16 [14]	21 to 70
EC4 [15]	20 to 50
AS 5100-6 [16]	25 to 65
ACI [17]	≥ 17.2

4. Failure modes

From the literatures it is found that the failure mode that affected the strength of stub and short members was the local buckling or yielding of the steel tube. Overall (half-sine) buckling affected the strength of long or slender members. Intermediate members failed by combined yield, overall buckling and/or local buckling. Buckling of steel tubes followed the failure of concrete core by shear and crushing. Failures like elephant foot buckling, drum-type, crippling in thin tubes were also observed. Weld crack was observed at the corners of certain specimens due to poor ductility of the weld. Hollow steel tubes and concrete-filled steel tubes showed a difference in the orientation of local buckling. Inward and outward buckling was observed in hollow steel tubes whereas CFST column failed by only by outward buckling due to the presence of concrete infill. Figure 2 shows the different failure patterns in CFST columns.

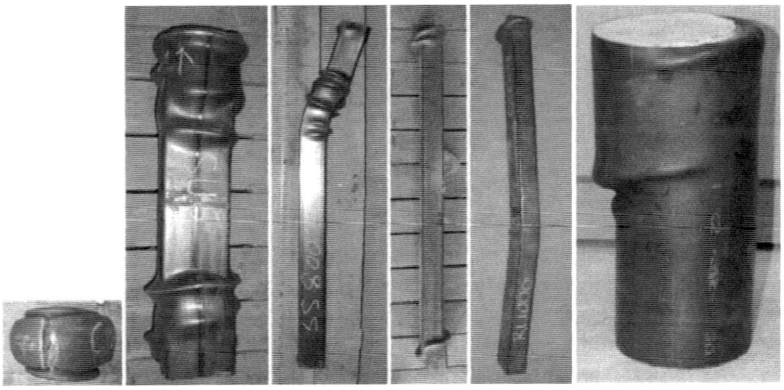

(a) (b) (c) (d) (e) (f)

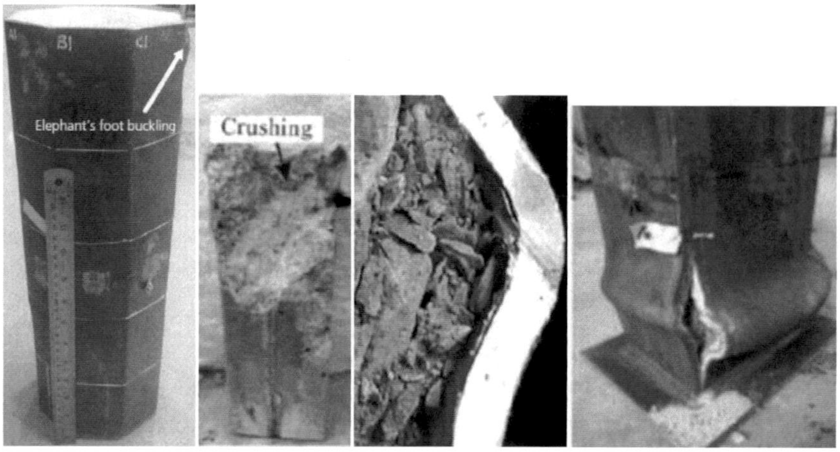

(g) (h) (i) (j)

Figure. 2 Failure patterns in CFST columns : (a) radial expansion [57], (b) local buckling [57], (c) plastic hinge [57], (d) shear [57], (e) global buckling [57], (f) crippling [27], Elephant's foot buckle [32], (h) concrete core crushing [54], (i) buckling of steel tube [54], (j) cracking of steel tube [19]

5. Parametric study

5.1. Concrete compressive strength (f_c)

Smaller confinement effect was observed in steel tubes filled with high strength concrete irrespective of the shape of the steel tube as the stiffness of the concrete increases with the concrete strength [25, 26, 27]. Increasing the value of compressive strength of concrete increased the bearing capacity of the CFST columns [3, 10, 23, 27, 28, 29]. High strength concrete core in CFST columns contributed to more compression load capacity for thinner tube [27]. Higher concrete strength improves the load bearing capacity of CFST columns especially those with smaller steel ratio [10]. Higher concrete compressive strength and strength of steel tube contributes to moderate improvement in the load carrying capacity of CFST columns filled with UHSC [30].

5.2. Steel tube thickness (t) and Steel yield strength (f_y)

The resistance of CFST columns increases with increase in steel yield strength [3, 7, 8, 13, 27]. The performance of concrete-filled steel tubular columns increases by increasing the thickness of steel tube [6, 10, 22, 26, 30].

5.3. Steel ratio (α)

Steel ratio is given by the ratio of area of steel to the total area of cross-section. For circular lightweight concrete-filled steel tube columns, low steel ratio showed better strength [32]. Improvement in load bearing capacity of CFST columns can be achieved with high steel ratio, low concrete grade and high strength steel [5, 7, 34].

5.4. Ductility

Ductility is the ability of a structure to maintain deformation beyond the elastic limit at the same time, maintaining a reasonable load carrying capacity until total failure [35]. The ductility index is influenced by the cross-section shape of the CFST column [32, 35, 36]. Ductility decreases in CFST columns filled with high strength concrete than steel tubes filled with low strength concrete [7, 8, 10, 13, 19, 28, 37, 39]. Stub columns have shown a higher ductility than that of the long columns. Higher ductility was observed in CFST columns that were shorter, bigger in diameter and filled with low strength concrete than columns with smaller diameters and high strength concrete infill [26]. Increase in steel ratio increases the ductility of CFST columns [8, 32, 34]. Increase in steel tube thickness offers better ductility to the CFST columns [6, 10, 27] and a bigger *B/D* ratio gives columns with weaker ductility [10]. High-strength steel in CFST columns demonstrates better ductility when lower D/t ratio and lower h/b ratio is used [18]. Ductility increases with increase in yield strength [27]. By applying the load on the concrete core only and by using by using Class 1 steel sections or by adopting higher steel contribution ratio, improvement in ductility and strength of UHSC filled composite columns can be observed [38]. The confinement index should not be taken > 3.0 for square ultra-high performance CFSTs [3]. An axial expansion of 5mm out of the steel tube by the lightweight concrete infill showed the ductile behavior of lightweight concrete-filled steel tubes [1].

5.5. Slenderness ratio
Greater flexibility, larger sideways displacement at the mid-height and lesser stiffness is observed in columns with large slenderness ratio [24, 29, 39].

5.6. Effect of steel fibers
Addition of steel fibers to UHSC increases the shear strength of concrete and the ultimate resistance of the column. Also, for the CFSTs with Class 2 or Class 3 sections, the ductility was improved compared to CFSTs with the plain UHSC [38].

5.7. Strength index
The strength index (SI) is the ratio between the calculated capacity of the column and the actual ultimate load on the column. Higher SI shows a stronger confinement of steel tube to concrete core and a delay in local buckling in the steel tube [30, 39]. SI of CFST columns increase with decrease in the D/t ratio [3]. Change in shape of cross-section decreases the SI [50]. The SI decreases as the width-to-thickness and aspect ratio increases and also the increase of confinement index (ξ) might not contribute to the increase of SI [19]. Slender columns had lower SI than short and intermediate length columns [27]. Increasing the confinement index does not give in high capacity for the rectangular CFST columns with high strength steel and results in the decrease of SI [19]. Use of high-strength steel in rectangular CFT columns showed high bearing capacity, low strength index and weaker confinement with increased D/t ratio and h/b ratio [18]. Parent hot-rolled circular concrete filled specimens showed larger SI values than the cold-formed elliptical CFST specimens [31].

5.8. Width to thickness ratio
The confinement given by the steel tube to concrete core decreased with bigger *B/D* ratio in round-ended rectangular stub CFST column specimens [10].

5.9. Concrete contribution ratio
Concrete contribution ratio (CCR) is defined as ratio of the ultimate load of the CFST column to the corresponding ultimate load of bare hollow steel tube. As the steel tube thickness decreases, CCR decreases [10]. CCR reduces with increase in width-to-thickness ratio and also with the increase of confinement index [19].

6. Axial load carrying capacity of CFST columns

6.1 Design equations from code of practices

Extensive research and the practical application of CFST construction has lead the way to develop design codes in many countries. Empirical formulae to predict the resistance of different cross-sections to compression from the codes of practice is shown in Table 4 (a) and (b).

ANSI/AISC 360 gives the formula to calculate the axial load carrying capacity of columns on the basis of compact, non-compact and, slender section that considers the critical buckling stress f_{cr}. For non-compact sections, the plastic strength P_p and the yield strength P_y of the composite section are considered. For compact sections, the compression capacity of the section is equal to the plastic strength of the composite section. The advantage of the effect of confinement provided by the circular steel tube is considered in the European code EC4 by applying an enhancement factor η_c and the reduction in the yield strength of the steel tube with diameter D, due to the lateral expansion of concrete is considered by applying the reduction factor η_a. The relative slenderness $\boldsymbol{\lambda}$, elastic critical force N_{cr}, effective flexural stiffness $(EI)_{eff}$ and the plastic resistance of the composite section to compressive normal force $N_{pl,Rk}$ are calculated as shown in Table 4. The Australian standard AS 5100-6 has similar formula as EC4 for calculating the compressive resistance of circular section under axial compression and uses the capacity factors $\emptyset$, $\emptyset_c$. A single equation is used in ACI 318-11 for calculating the compressive load carrying capacity of circular and rectangular sections. In the following equations, area of steel, area of concrete, strength of steel, strength of concrete, second moments of area of the steel and the concrete are denoted by A_s, A_c, f_y, f_c, I_s and I_a respectively.

Table 4. Design equations from codes for compressive load carrying capacity of CFST columns.

4(a)

Code	Cross- section shape	Equation for axial load (N_u) carrying capacity of CFST column
ANSI/AISC 360 (2005) [14]	Circular Compact	$N_u = A_s f_y + 0.95 A_c f_c$
	Non- compact	$N_u = P_p - [P_p - P_y/(\lambda_r - \lambda_p)^2](\lambda - \lambda_p)^2$ $\lambda_p = 0.15 E_s/f_y$; $\lambda_r = 0.19 E_s/f_y$ $P_y = A_s f_y + 0.7 A_c f_c$
	Slender	$N_u = A_s f_{cr} + 0.7 A_c f_c$ $f_{cr} = 0.72 f_y \left[\left(\left(\frac{D}{t}\right)\frac{f_y}{E_s}\right)^{0.2}\right]^{-1}$
	Rectangular Compact	$N_u = A_s f_y + 0.85 A_c f_c$
	Non- compact	$N_u = P_p - [P_p - P_y/(\lambda_r - \lambda_p)^2](\lambda - \lambda_p)^2$ $P_y = A_s f_y + 0.7 A_c f_c$

4(b)

ANSI/AISC 360 (2005) [14]	Rectangular Slender	$N_u = A_s f_{cr} + 0.7 A_c f_c$ $f_{cr} = 0.9 E_s (B/t)^{-2}$
EN1994-1-1 (2004) [15]	Any type of section	$N_u = A_s f_y + A_c f_c$
	Circular	$N_u = \eta_a A_s f_y + \left[1 + \eta_c \left(\frac{t}{D}\right)\left(\frac{f_y}{f_c}\right)\right] + A_c f_c$ $\eta_a = 0.25(3 + 2\lambda) \leq 1.0$ $\eta_c = 4.9 - 18.5\lambda + 17\lambda^2 \geq 0$ $\lambda = \left(\frac{N_{pl,Rk}}{N_{cr}}\right)^{1/2}$, $N_{pl,Rk} = A_s f_y + 0.85 A_c f_c$ $N_{cr} = \pi^2 (EI)_{eff}/L^2$; $(EI)_{eff} = E_a I_a + E_s I_s$
AS 5100-6 (2004) [16]	Rectangular	$N_u = \emptyset A_s f_y + \emptyset_c A_c f'_c$ $\emptyset = 0.9, \emptyset_c = 0.6$
	Circular	$N_u = \emptyset A_s \eta_a f_y + \emptyset_c A_c f_c \left(1 + \frac{\eta_c t f_y}{D f_c}\right)$
ACI 318-11 (2011) [17]	Circular and Rectangular	$N_u = A_s f_y + 0.85 A_c f_c$

6.2. Validation of the design equations from codes

EC4 gave reasonable predictions for axial load carrying capacity of circular CFST columns [1, 25, 26, 27, 51, 52, 53]. Also, it is suggested by Xiong et al. [38] that for circular CFST columns filled with UHSC or when Class 3 sections are not used, the confinement effect in EC4 should be ignored. Calculated results using design equations in EC4 were closer to experimental results of rectangular and square [26, 35, 27, 58], elliptical [6, 22, 24], hexagonal [7, 32], octagonal [32], triangular [32] and special-shaped [50] CFST columns subjected to axial compression. The South African code SANS 10162-1 and the Japanese code AIJ also gave satisfactory predictions for axially loaded circular CFST columns [26, 54].

The Chinese code DBJ /T13-51-2010 can be used to compute the compression strengths of square and rectangular [52, 54], hexagonal [7] and special-shaped [50] CFST columns. The compression strength of rectangular CFST columns can be calculated using the codes ACI 318 [54, 57] and GJB 4142-2000 [54]. The code ANSI/AISC 360 predicts well for circular [27] and elliptical [55] CFST columns under axial compression.

6.3. Design equations modified and proposed

Apart from the equations given in codes, equations were also proposed by researchers. A summary of the equations proposed for concrete-filled steel tubes of different cross-sectional shapes are shown in Table 5 (a), (b) and (c). For circular stub CFST columns, Lu and Zhao [41] proposed modified equations for the design codes AIJ (Japanese), ACI, AS, AISC, EC4, DL/T (Chinese) and Han et al. [59]. Equations were proposed by Ding et al. [40] to predict the ultimate capacity (N_u) or ultimate strength ($f_{sc,u}$) of centrally loaded circular CFT stub column, by Xue et al. [43] to calculate the ultimate load capacity of circular CFST stub columns with de-bonding (N_D) given as the product of ultimate load capacity of specimens without debonding (N_u) and debonding reduction factor (K_D), by Wang et al. [45] for the axial bearing capacity of circular CFT that considered the size effect. For lightweight concrete-filled steel tube columns, equations were proposed by Zhongqiu et al. [42] and Fu et al. [44] by considering a coefficient of lateral deformation $\boldsymbol{\nu}$ and a lateral pressure coefficient k.

Using the principle of superposition, new equation was proposed by Ding et al. [28] for calculating the ultimate bearing capacity axially-loaded square CFT stub columns with a confinement factor 1.2, Ding et al. [13] for ultimate strength of hexagonal CFT stub column with a confinement factor 1.3, Ding et al. [8] for the ultimate load bearing capacity of axially loaded octagonal CFT stub columns with a confinement factor of 1.5.

The axial compressive bearing capacity of the square ultra-high performance CFST short columns was proposed by Yan et al. [30]. The limitations in DB 29-57 was extended using a new method proposed by Du et al. [19] to include steel grade 460 MPa in designing rectangular CFT columns. The axial load carrying capacity of the elliptical CFST stub columns was proposed by Uenaka [46] and the re-estimations (N_{rest-a}, N_{rest-b}) were calculated with concrete strength induced by two confined stress in larger and smaller directions ${}_{c}\sigma_{cBa}$, ${}_{c}\sigma_{cBb}$ respectively. To design regular hexagonal CFST short columns in compression with internal angle $\theta = 120^{\circ}$, an equation was proposed by Hassanein et al. [3]. A new formula for lateral confinement pressure f_{rp} was also proposed. Equations for round-ended rectangular CFST stub and short columns were proposed by Faxing et al. [10] and Hassanein and Patel [5] respectively. Unified formulas to predict the axial load bearing capacity for circular hollow and solid concrete-filled steel tube columns, and for circle and polygon CFST columns under axial compression were proposed by Yu et al. [47] and Yu et al. [48] respectively. A simple superposition model to predict the ultimate strength for both circular and rectangular CFST stub columns under axial compression was given by Wang et al. [49].

Table 5. Equations proposed by various researchers for different cross-sectional shapes.

5(a)

Cross-section	Author	Year	Proposed equation on ultimate bearing capacity of axially loaded columns (N_u)
Circular	Lu and Zhao [41]	2010	$N_u = {f_{cyl,100}}^{a} A_c + 1.4 f_y A_s$ (for AIJ) $N_u = {f_{cyl,150}}^{b} A_c + 1.47 f_y A_s$ (for ACI,AISC,AS) $N_u = \left(1 + 1.8 \frac{t}{D} \frac{f_y}{f_{cyl,150}}\right) f_{cyl,150} + f_y A_s$ (for EC4) $N_u = (1.3 + 1.1\xi^{c}) f_{ck} (A_s + A_c)$ (for DL/T and Han et al.[59])
	Ding et al.[40]	2011	$N_u = f_c A_c (1 + 1.7\xi)$, $f_{sc,u} = (1 + \rho)(1 + 1.7\xi) f_c$, $\xi = \alpha \frac{A_s}{A_c}$
	Zhongqiu et al. [42]	2011	$N_u = 1.35(f_y A_s + f_c A_c)$
	Xue et al.[43]	2012	$N_D = K_D N_u$ $K_D = \begin{cases} 1 - a_1 R_d & (0 \le R_d \le 0.3) \\ 1.015 - 0.3\, a_1 - 0.05 R_d & (0.3 \le R_d \le 1) \end{cases}$ $a_1 = \begin{cases} 0.25\xi + 0.25 & (0.6 \le \xi \le 1.5) \\ 0.625 & (1.5 \le \xi) \end{cases}$
	Fu et al. [44]	2015	$N_u = f_y A_s + f_c A_c + (2kt\nu f_y / D)$, $\nu = 0.5$, k = 3.4
	Wang et al.[45]	2016	$N_u = A_s f_{sz} + A_c f_{cc}$ Strength of the confined concrete $f_{cc} = f_{cD} + K^{d} f_r$

			Size-dependant concrete compressive strength $f_{cD} = 1.073Rf_c$ Confining stress provided by steel tube $f_r = \frac{-2t}{D-2t}f_{s\theta}$ Hoop stress in the steel tube $f_{s\theta} = G_\theta f_y$ Vertical stress in the steel tube $f_{sz} = -G_z f_y$ Hoop stress factor $G_z = 0.32D^{0.12}$ Vertical stress factor $G_\theta = 2.76D^{-0.29}$

[a]$f_{cyl,100}$,[b]$f_{cyl,150}$ – concrete cylindrical compressive strength in MPa.
[c]ξ- confinement factor.
[d]K –varies with G_θ.

5(b)

Rectangular/ Square	Ding et al. [28]	2014	$N_u = A_cf_c + 1.2A_sf_s$
	Du et al. [19]	2016	$N_u = f_sA_s + (1+k)f_cA_c$ Augmentation factor $k = 0.5668 - 0.0039h/t\sqrt{f_y/235}$
	Yan et al. [30]	2019	$N_u = A_s(0.89f_y) + A_c\left\{\left[1 + 3.5 \text{ x } \left(\frac{\left(\frac{1}{3} \text{ x } \frac{2t}{b-2t}\right) \text{ x } 0.19f_y}{f_c}\right)^{0.72}\right] f_c\right\}$
Elliptical	Uenaka [46]	2014	$N_u = 1.46(A_cf'_c + A_sf_y)$ $N_{rest-a} = A_{c_c}\sigma_{cBa} + A_sf_y$ $N_{rest-b} = A_{c_c}\sigma_{cBb} + A_sf_y$ ${}_c\sigma_{cBa} = f'_c + k\frac{t}{a-t}\sigma_{\theta a}$ ${}_c\sigma_{cBb} = f'_c + k\frac{t}{b-t}\sigma_{\theta b}$
Hexagon (Regular)	Ding et al. [13]	2016	$N_u = f_cA_c + 1.3f_yA_s$, $A_c = \frac{3\sqrt{3}b^2}{2}$, $A_s = 6bt$
	Hassanein et al. [3]	2017	$N_u = (\gamma_c f'_c + 4.1f_{rp})A_c + f_yA_s$ $f_{rp} = \begin{cases} \left(0.0491703 - 0.0007943\frac{B+D}{2t}\right)f_{sy} \text{ for } 17 \leq \frac{B^e+D^f}{2t} < 63 \\ \left(0.0065311 - 0.0000044\frac{B+D}{2t}\right)f_{sy} \text{ for } 17 \leq \frac{B+D}{2t} \leq 63 \end{cases}$
Octagon	Ding et al. [8]	2016	$N_u = f_cA_c + 1.5f_yA_s$

Round-Rectangular	Faxing et al. [10]	2015	$N_u = A_c f_c[1 + (0.8 + 0.9D/B)\xi]$
	Hassanein and Patel [5]	2018	$N_u = \gamma_s f_y A_{s,RE}{}^{g} + \left(\gamma_c f'_c + 4.1 f'_{rp}\right) A_{C,RE}{}^{H} + f_y A_{S,RP,eff}{}^{i} + \gamma_c f'_c A_{C,RP}{}^{j}$ $\gamma_s = 1.458(D/t)^{-0.1}\ (0.9 \le \gamma_s \le 1.10)$ Confinement pressure $f'_{rp} = \begin{cases} 0.7(\nu_e{}^{k} - \nu_s{}^{l})\left[\frac{2t}{D-2t}\right] f_{sy} & \text{for } \frac{D}{t} \le 47 \\ \left(0.006241 - 0.0000357\frac{D}{t}\right) f_{sy} & \text{for } \frac{D}{t} \le 150 \end{cases}$

[e]B, [f]D – Thickness and length of hexagon, width and thickness of round-rectangular column.
[g]$A_{s,RE}$ - steel tube cross-sectional area in the round-ended column.
[h]$A_{c,RE}$ - concrete cross-sectional area in the round-ended column.
[i]$A_{s,RP,eff}$ - steel effective cross-sectional area in the rectangular shape of the round-ended column.
[j]$A_{C,RP}$ – concrete cross- sectional area in the rectangular shape of the round-ended column.
[k]v_e- Poisson's ratio of steel tube filled with concrete.
[l]v_s- Poisson's ratio of steel tube devoid of concrete.

5(c)

Solid and hollow CFST	Yu et al. [47]	2010	$N_u = \left(1 + 0.5k_n{}^{i}\frac{\xi}{1+\xi}\right)(f_y A_s + f_c A_c)$, $f_{sc} = (1 + \eta_c)[(1 - \beta)f_c + \beta f_y]$ $\eta_c = \Omega\xi_{sc}/\left(\left[2.0\Omega + 0.05\xi_{sc} + \left(0.2\frac{f_{ck}}{f_y} - 0.05\right)\xi_{sc}\Omega\right](\Omega + \xi_{sc})\right)^{1/2}$ Steel area ratio $\beta = \mathrm{A_s}/(\mathrm{A_s} + \mathrm{A_c})$, solid ratio $\Omega = \mathrm{A_c}/(\mathrm{A_c} + \mathrm{A_k})$
Circle and polygon	Yu et al.[48]	2013	$N_u = \varphi_{sc} N_0$ $\varphi_{sc}{}^{j} = \frac{1}{2\bar{\lambda}_{sc}{}^{2}}\left[\bar{\lambda}_{sc}{}^{2} + K\bar{\lambda}_{sc} + 1 - \sqrt{\left(\bar{\lambda}_{sc}{}^{2} + K\bar{\lambda}_{sc} + 1\right)^2 - 4\bar{\lambda}_{sc}{}^{2}}\right]$ $K = 0.25 - 0.09k_e$ Normalized slenderness ratio $\bar{\lambda}_{sc} = \frac{L_o{}^{k}}{\pi}\left(\frac{N_0}{E_{sc} I_{sc}}\right)^{1/2}$ $E_{sc} I_{sc} = E_c I_c + E_s I_s$. , $N_0 = (1 + \eta^{l})(f_y A_s + f_c A_c)$ $\eta = 0.5k_e\frac{\xi}{1+\xi}$, $\xi = A_s f_y / A_c f_{ck}$ confinement effectiveness coefficient (k_e) $k_e = (1 - \psi^m)(n^2 - 4)/(1 - \psi)(n^2 - 4)$, $\psi = 1 - \Omega$
Circle and Rectangle	Wang et al. [49]	2017	$N_u = N_a + N_c = \phi f_y A_s + \phi_c f'_c A_c$ Capacity factors: $\phi = 0.83$, $\phi_c = 0.78$

[i]k_n- polygon confinement effectiveness coefficient.
[j]φ_{sc}- Stability factor.
[k]L_o- effective length of column.
[l]η - enhanced confining coefficient.

$^{m}\psi$ – hollow ratio.

7. Conclusion

The circular hollow steel sections show enhanced confinement to the concrete core than any other cross-sectional shaped hollow sections. Cross-sectional shapes like rectangular, square, hexagonal, octagonal and other special-shaped CFST columns have the strongest confinement at their corners and at the center. The round-ended CFST column shows the strongest confinement in the round ends. Large size CFST columns show weaker confinement effect. Composite action is stronger in slender CFST columns than stocky ones. Confinement is effective in steel tubes filled with normal strength concrete than high strength concrete-filled steel tubes. On the other hand, bearing capacity of the CFST columns increases with increase in compressive strength of concrete. High strength concrete in thicker tubes with high yield strength offers greater confinement effect. Codes should revise their limits on concrete compressive strength and steel yield strength due to the emerging use of high strength and ultra-high strength materials in construction.

CFST columns with lightweight concrete infill are highly prone to local buckling than CFST columns with normal strength concrete infill. The failure mode of CFST columns is not influenced by the aggregate type used in the infill concrete. When lightweight concrete-filled steel tube columns are used in structures, the design confinement factor should not be less than 1.0 to provide enough constraint to the concrete. When ultra-high strength concrete is used in CFST, a careful match of the steel tube grade to the concrete strength should be done. The bond strength between the steel tube and ultra-high performance core concrete increases with the use of an expansive agent.

The ultimate load capacity values increases with the increment of steel wall thickness. Thicker tubes offer greater confinement which enhances the resistance of the composite section. Specimens with pointed ends have lower ductility than round specimens. Ductility decreases with increase in steel grade. Ductility in ultra-high strength concrete filled steel tubes can be improved by adding a minimum of 1% of steel fibers. Slenderness ratio should be small to achieve greater stiffness in CFST columns. EC4 is found to give closer predictions to experimental results of CFST columns of different shapes considered in this paper.

References

[1] Hamidian, M R, Shafigh P, Jumaat, M Z, Alengaram, U J and Sulong N R 2016 A new sustainable composite column using an agricultural solid waste as aggregate *J.Clean Prod.* **129** 282-91
[2] Chan T M and Gardner L 2008 Compressive resistance of hot-rolled elliptical hollow sections *Eng. Struct.* **30(2)** 522-32
[3] Hassanein M F Patel V I and Bock M 2017 Behaviour and design of hexagonal concrete-filled steel tubular short columns under axial compression *Eng. Struct.* **153** 732-48
[4] Zhu J Y Chan T M and Young B 2019 Cross-sectional capacity of octagonal tubular steel stub columns under uniaxial compression *Eng. Struct.* **184** 480-94
[5] Hassanein M F and Patel V I 2018 Round-ended rectangular concrete-filled steel tubular short columns: FE investigation under axial compression *J. Constr. Steel Res.* **140** 222-36
[6] Yang H Lam D and Gardner L 2008 Testing and analysis of concrete-filled elliptical hollow sections *Eng. Struct.* **30(12)** 3771-81
[7] Xu W Han L H and Li W 2016 Performance of hexagonal CFST members under axial compression and bending *J. Constr. Steel Res.* **123** 162-75
[8] Ding F X Li Z Cheng S and Yu Z W 2016 Composite action of octagonal concrete-filled steel tubular stub columns under axial loading *Thin-Walled Struct.* **107** 453-61

[9] Zhu J Y and Chan T M 2018 Experimental investigation on octagonal concrete filled steel stub columns under uniaxial compression *J. Constr. Steel Res.* **147** 457-67
[10] Faxing D Lei F Zhiwu Y and Gang L 2015 Mechanical performances of concrete-filled steel tubular stub columns with round ends under axial loading *Thin-Walled Struct.* **97** 22-34
[11] Wang F C and Han L H 2018 Analytical behavior of special-shaped CFST stub columns under axial compression *Thin-Walled Struct.* **129** 404-17
[12] Lu Z H Zhao Y G and Yu Z W 2011 A Strength Model for Square CFT Stub Columns with Compact Sections In *Appl. Mech. Mater.* **94** 425-30
[13] Ding F X Li Z Cheng S and Yu Z W 2016 Composite action of hexagonal concrete-filled steel tubular stub columns under axial loading *Thin-Walled Struct.* **107** 502-13
[14] American Institute of Steel Construction ANSI/AISC 360: Specification for Structural Steel Buildings. Chicago Illinois 2005
[15] Eurocode 4 Design of composite steel and concrete structures Part 1-1: general rules and rules for buildings London: British Standards Institution 2004
[16] Standards Australia Bridge Design Part 6: Steel and Composite Construction AS5100.6-2004 2004 Sydney Australia
[17] American Concrete Institute Building Code Requirements for Structural Concrete and Commentary ACI 318-11 2011 Farmington Hills MI
[18] Du Y Chen Z and Yu Y 2016 Behavior of rectangular concrete-filled high-strength steel tubular columns with different aspect ratio *Thin-Walled Struct.* **109** 304-18
[19] Du Y Chen Z and Xiong M X 2016 Experimental behavior and design method of rectangular concrete-filled tubular columns using Q460 high-strength steel *Constr Build Mater.* **125** 856-72
[20] Gardner L and Chan T M 2007 Cross-section classification of elliptical hollow sections *Steel Compos. Struct.* **7(3)** 185
[21] Ruiz-Teran A M and Gardner L 2008 Elastic buckling of elliptical tubes *Thin-Walled Struct.* **46(11)** 1304-18
[22] Zhao X L and Packer J A 2009 Tests and design of concrete-filled elliptical hollow section stub columns *Thin-Walled Struct.* **47(6-7)** 617-28
[23] Chen J Zhu J Y and Chan T M 2020 Experimental and numerical investigation on stub column behaviour of cold-formed octagonal hollow sections *Eng. Struct.* **214** 110669
[24] Mahgub M Ashour A Lam D and Dai X 2017 Tests of self-compacting concrete filled elliptical steel tube columns *Thin-Walled Struct.* **110** 27-34
[25] Ghannam S Al-Rawi O and El-Khatieb 2011 Experimental study on light weight concrete-filled steel tubes M *Jordan J. Civ. Eng.* **159(696)** 1-9
[26] Dundu M 2012 Compressive strength of circular concrete filled steel tube columns *Thin Walled Struct,* **56** 62-70
[27] Ekmekyapar T and Al-Eliwi B J 2016 Experimental behaviour of circular concrete filled steel tube columns and design specifications *Thin-Walled Struct.* **105** 220-30
[28] Ding F X Fang C Bai Y and Gong Y Z *J 2014* Mechanical performance of stirrup-confined concrete-filled steel tubular stub columns under axial loading *Constr. Steel Res.* **98** 146-57
[29] Evirgen B Tuncan A and Taskin 2014 Structural behavior of concrete filled steel tubular sections (CFT/CFSt) under axial compression *Thin-Walled Struct.* **80** 46-56
[30] Yanv Y Xu L Li B Chi Y Yu M Zhou K and Song 2019 Axial behavior of ultra-high performance concrete (UHPC) filled stocky steel tubes with square sections *J. Constr. Steel Res.* **158** 417-28
[31] Chan T M Huai Y M and Wang W 2015 Experimental investigation on lightweight concrete-filled cold-formed elliptical hollow section stub columns *J. Constr. Steel Res.* **115** 434-44

[32] Zhu J Y and Chan T M 2018 Behaviour of polygonal-shaped steel-tube columns filled with high-strength concrete Experimental study on ultrahigh strength concrete filled steel tube short columns under axial load *Proceedings of the Institution of Civil Engineers-Structures and Buildings* **171(2)** 96-112
[33] Zhou X Mou T Tang H and Fan B 2017 Experimental study on ultrahigh strength concrete filled steel tube short columns under axial load *Adv. Mater. Sci. Eng.* 2017
[34] Song T Y and Xiang K 2020 Performance of axially-loaded concrete-filled steel tubular circular columns using ultra-high strength concrete In *Struct.* **24** 163-76
[35] Yu X Tao Z and Song TY 2016 Effect of different types of aggregates on the performance of concrete-filled steel tubular stub columns *Mater Struct* **49(9)** 3591-605
[36] Al Mamoori A H N Naser F H and Dhahir M K 2020 Effect of Section Shape on the Behaviour of Thin Walled Steel Columns Filled with Light Weight Aggregate Concrete: Experimental Investigation *Case Stud. Constr. Mater.* p.e 00356
[37] Li Z B Xie F Wang W J and Ma H 2015 The correction coefficient of size effect on the axial compression strength of circular concrete-filled steel tubular short columns *Mater. Res. Innov.* **19(sup1)** S1-177
[38] Xiong M X Xiong D X and Liew J R 2017 Axial performance of short concrete filled steel tubes with high-and ultra-high-strength materials *Eng. Struct.* **136** 494-510
[39] Jamaluddin N Lam D Dai X H and Ye J 2013 An experimental study on elliptical concrete filled columns under axial compression *J. Constr. Steel Res.* **87** 6-16
[40] Ding F X Yu Z W Bai Y and Gong Y Z 2011 Elasto-plastic analysis of circular concrete-filled steel tube stub columns *J. Constr. Steel Res.* **67(10)** 1567-77
[41] Lu Z H and Zhao YG 2010 Suggested empirical models for the axial capacity of circular CFT stub columns *J. Constr. Steel Res.* **66(6)** 850-62
[42] Zhongqiu F Bohai J Yu Z and Xiaoliang W 2011 An experimental behavior of lightweight aggregate concrete filled steel tubular stub under axial compression *Design, Construction, Rehabilitation, and Maintenance of Bridges* 24-32
[43] Xue J Q Briseghella B and Chen BC 2012 Effects of debonding on circular CFST stub columns *J. Constr. Steel Res.* **69(1)** 64-76
[44] Fu Z Q Ji B H Ge H B and Yang M 2015 Combined strength of lightweight aggregate concrete-filled steel tube *Mater. Res. Innov.* **19(sup5)** S5-898
[45] Wang W Tang Z Li Z and Ma H 2016 Bearing capacities of different-diameter concrete-filled steel tubes under axial compression *Adv. Mater. Sci. Eng.* 2016
[46] Uenaka K 2014 Experimental study on concrete filled elliptical/oval steel tubular stub columns under compression *Thin-Walled Struct.* **78** 131-7
[47] Yu M Zha X Ye J and She C 2010 A unified formulation for hollow and solid concrete-filled steel tube columns under axial compression *Eng. Struct.* **32(4)**1046-53
[48] Yu M Zha X Ye J and Li Y 2013 A unified formulation for circle and polygon concrete-filled steel tube columns under axial compression *Eng. Struct.* **49**1-10
[49] Wang Z B Tao Z Han LH Uy B Lam D and Kang W H 2017 Strength, stiffness and ductility of concrete-filled steel columns under axial compression *Eng. Struct.* **135** 209-221
[50] Ren Q X Han L H Lam D and Hou C 2014 Experiments on special-shaped CFST stub columns under axial compression *J. Constr. Steel Res.* **98** 123-33
[51] Zhu L Ma L Bai Y Li S Song Q Wei Y Zhang L Zhang Z and Sha X 2016 Large diameter concrete-filled high strength steel tubular stub columns under compression *Thin-Walled Struct.* **108** 12-19
[52] Zhou S Sun Q and Wu X 2018 Impact of D/t ratio on circular concrete-filled high-strength steel tubular stub columns under axial compression *Thin-Walled Struct.* **132** 461-74

[53] Huang W Fan Z Shen P Lu L and Zhou Z *2020* Experimental and numerical study on the compressive behavior of micro-expansive ultra-high-performance concrete-filled steel tube columns *Constr Build Mater.* **254** 119150
[54] Zhang X Kuang X Yang J and Fan Y 2019 Bearing capacity of stone-lightweight aggregate concrete-filled steel tubular stub column subjected to axial compression *KSCE J Civ Eng* **23(7)** 3122-34
[55] Lam D and Testo N 2011 Structural design of concrete filled steel elliptical hollow sections In *Steel Compos. Struct.* **VI** 252-62
[56] Hossain K M and Chu K 2019 Confinement of six different concretes in CFST columns having different shapes and slenderness *Int. J. Adv. Struct. Eng.* ***11*(2)** 255-70
[57] Thumrongvut J and Tiwjantuk P 2018 Strength and axial behavior of cellular lightweight concrete-filled steel rectangular tube columns under axial compression *In Mater. Sci. Forum* **941** 2417-22
[58] Thumrongvut J Seangatith S Siriparinyanan T and Wangrakklang S 2016 An experimental behaviour of cellular lightweight concrete-filled steel square tube columns under axial compression *In Mater. Sci. Forum* **860** 121-4
[59] Han L H Yao G F Zhao X L 2005 Experiment behavior of thin-walled hollow structural steel (HSS) stub columns filled with self-consolidating concrete (SCC) *Journal J. Constr. Steel Res.* **61(9)** 1241-69

Defluoridation of Water by Using Modified Material Developed from *Ficus benghalensis* Leaf: Characterization, Kinetic and Thermodynamic Study

Nandkishor G. Telkapalliwar

Department of Chemistry, Dr Ambedkar College, Deeksha Bhoomi, Nagpur, India

Corresponding author e-mail: telkapalliwar80@gmail.com

Abstract. In the present study, microwave-assisted and chemically treated carbonized *Ficus benghalensis* leaf (MACFBL) material was used as an adsorbent material for the removal of fluoride from water. The results of the characterization of MAFBL carbon material exhibited ideal adsorbent properties. The fluoride adsorption experiments were performed under the batch mode to improve the different affecting parameters such as contact time (0-300 min) and temperature (303-343K) at predetermined pH (5), agitation speed (120 strokes/min), fluoride concentration (2 mg/L) and adsorbent dose (5 g/L). The maximum fluoride removal efficiency of fluoride on MACFBL material was found 86.5 %. The fluoride adsorption data applied for four well known kinetic models such as pseudo first-order, pseudo second-order, intra-particle diffusion, and Elovich kinetic models. The pseudo second-order kinetic study shows the most favourable mechanism for the removal of fluoride. Thermodynamic investigation proposed that the fluoride adsorption process onto MACFBL was exothermic. The instrumental study of MACFBL adsorbent material before and after adsorption during FTIR,SEM, EDX and XRD techniques established the fluoride adsorption on the carbon surface. The developed adsorbent material (MACFBL) is efficient for the removal of fluoride.

1. Introduction

The rise in high concentrations of fluoride in groundwater is of significant concern worldwide due to natural and anthropogenic actions contributing to health threats for the human population [1,2]. In general, fluoride in ground is derived through the degradation of distinctive natural minerals found in the different soils and rocks with which the water interacts[3]. Fluoride can, other than natural sources, originate in effluents from the semiconductor, metal processing, fertilizer and glass manufacturing industries [4,5]. The maximum allowable limit of fluoride content in drinking water for a human being is 1.5 mg / L, as set by the Bureau of Indian Standards (BIS) and WorldHealth Organization (WHO) [6,7]. Fluoride is an extremely toxic ion at concentrations greater than the allowable level and may contribute to multiple health issues, such as skeletal fluorosis, dental fluorosis, neurological injury, osteoporosis, parathyroid damages, Alzheimer's disease, male infertility, thyroid and kidney problems, liver damage, etc.[8].

In many developed and developing countries, including Asia, the USA, and Africa, higher fluoride concentrations in groundwater have been recorded. It has been found that with excess fluoride concentration, 17 different states of India are strongly affected, particularly Andhrapradesh, Telangana, Rajasthan, Madhya Pradesh, Gujarat, Assam, Uttar Pradesh and Tamilnadu [9, 10]. Owing to their adverse effects on humanwellbeing and the environment, various methods for the expulsion of fluoride from water have been developed over a few decades. That includes ultrafiltration, separation of solvents, sedimentation, absorption, coagulation, reverse osmosis, electrodialysis, ion exchange, adsorption, etc. [11, 12, 13]. In view of its low-cost installation, quick and simple maintenance, effective efficiency and fast operating procedure, adsorption is extensively considered and adopted among all these methods [12, 13].

The condition of excess fluoride content in drinking water is increasing gradually, and commonly used adsorption methods have been used to conduct water defluoridation. In recent years, activated carbons and various unconventional biomass materials such as Citrus limonum (lemon) leaf [3], KOH-treated Syzygium cumini Seed [10], Sawdust [11], tree bark [14], leaves [15], fruit shell / peal [16] and fruit seed [17] have been investigated and recorded for fluoride removal due to their low cost, simple availability in large amounts and effective fluoride removal. Industrial activated carbon treated with sugarcane bagasse[18], Hyacinth beads [19], aluminium impregnated coconut fibre [20], Camellia oleifera seed shell,[21], Santalum Album Leaf Powder[22], Al-impregnated Eucalyptus bark[23], Tea waste biomass[12, 24] and various carbon material identified by researchers for defluoridation of water [25]. The purpose of the present study is to set up and investigate the kinetic and thermodynamic efficacy of non-conventional adsorbent material obtained for the removal of fluoride from the aqueous solution using Ficus benghalensis leaves.

2. Material and Method

Analytical grades chemicals were used in the present work, such as sodium hydroxide, acetic acid, hydrochloric acid, nitric acid, sulphuric acid, sodium fluoride, sodium bicarbonate, sodium sulphate, sodium chloride, sodium phosphate, potassium hydroxide, sodium nitrate, 1, 2-cyclohexane diaminetetraacetic acid (CDTA) and these chemicals were acquired from S-D Fine Chemicals Ltd or Merck India Limite Ltd.

2.1. Preparation of Adsorbent

The *Ficus benghalensis* leaf collection was taken from the neighbourhood area. To extract dust and additional impurities, it was washed with water. It was ground using a home blender after air drying and then sieved through a mesh size of 300. The leaf samplewas rinsed with double distilledwater and was dried for 24 hours in an oven at 80 ^{O}C. The resulting dried leaf powder of *Ficus benghalensis* was carbonised in the muffle furnace at 500°C for 5 hours. This carbonised leaf powder was again triggered for 30 minutes at one-minute intervals in a domestic microwave (900MW). The microwave-assisted carbonised leaf powder was then individually impregnated for 24 hours with 0.5 N sodium hydroxide and 0.5 N sulfuric acid. With double purified water, the resulting carbon material was washed until a stable pH of the slurry was produced. Finally, the carbon sample was dried in a vacuum oven at 110 ^{O}C, mashed well and kept for future use in airtight plastic containers. Microwave-assisted carbonised *Ficus benghalensis* leaf is the activated material developed from the *Ficus benghalensis* leaves and was referred to as MACFBL subsequently (Figure 1).

2.2. Characterization of adsorbent

2.2.1. Proximate analysis

Physico-chemical parameters of MACFBL were studied, like pH, moisture, ash, volatile matter, bulk density, fixed carbon, water, and acid-soluble matter of MACFBL were examined. The results of the ultimate analysis were summarized in Table 1. Using the Elico pH metre, model LI-120, the pH of the

MACFBL bio adsorbent material was determined and other parameters were evaluated by utilizing standard analytical test methods [28, 29, 30].

Leaves *of Ficus benghalensis* | Dry crushed leaves *Ficus benghalensis* | Untreated leaf powder of *Ficus benghalensis* | *Ficus benghalensis* Activated Carbon **(MACFBL)**

Figure 1: Graphical representation of the development of carbon material (MAFBL)

2.2.2. Ultimate analysis.

The Brunauer-Emmett-Teller (BET) surface area pore characteristics were analyzed by using by a Quanta Chrome Nova-1000 surface analyzer instrument using a computer and software-controlled nitrogen gas adsorption analyzer at -196^{0}C. Using the Elementar Vario EL III model (C-H-N Analyser), the elements like C, H, N, and S were analysed. Scanning Electron Microscopy (SEM) (JEOL Model JSM-6390LV) was used for the study of MACFBL's microstructure morphology. Electron dispersive X-ray (EDX) (JEOL JSM-7600F FEG-SEM model) was used for the element and chemical classification of the developed MACFBL material. The FTIR spectra wereobtained by using Fourier-Transform-Infrared-Spectrophotometer (Thermo Nicolet, Avatar 370) with KBr. The FTIR spectra give information about the characteristic functional groupsnearby on the MACFBL bio-adsorbent material. The Structural integrity of the bio-adsorbent samples was checked by Powder X-ray diffraction (Model: Bruker AXS D8 Advance diffractometer).

2.3. Batch adsorption experiments

The developed MACFBL carbon material was used for the defluoridation of water by batch adsorption experiments at different initial concentrations of fluoride (2 mg/L to 10 mg/L). The 50 ml of known synthetic fluoride concentration solutions were taken for a batch test in 100 ml of Erlenmeyer flask and were shaken at 120 strokes/min for prearranged contact time, adsorbent dose, temperature and pH. The fluoride concentrations before and after adsorption were estimated by utilizing fluoride ion-selective electrode (HANNAModel No. H I 4110) and ion-selective meter (HANNA Model No. HI 4522). The adsorption capacities of fluoride were determined by the equation(1):

$$q_e = \frac{(Co - Ce)\, V}{m} \qquad (1)$$

Where m, V, C_0 , C_e and q_e are the mass of adsorbent (g), the volumeof the sample solution (L), initial concentration of fluoride (mg/L), fluoride concentration at equilibrium (mg/L) and amount of fluoride adsorbed at equilibrium (mg/g), respectively. The fluoride removal efficiency from the water was evaluated by equation (2):

$$\%\ \text{Removal of fluoride} = \frac{(Co - Ce)}{Co}\ X\ 100 \qquad (2)$$

The impacts of agitation speed, adsorbent dose, contact time, pH, temperature and initial fluoride concentration have been considered for fluorideremoval from the aqueous solutions by

IOP Conf. Series: Materials Science and Engineering **983** (2020) 012008 doi:10.1088/1757-899X/983/1/012008

utilizing MACFBL material. The kinetic of fluorideadsorption was studied by applying four different models, e.g., pseudo first-order, pseudo second-order, Elovich and Intra-particle diffusion models[**Table 1**]. The thermodynamic analysis for enthalpychange,entropychange andfree energychangewas carried outthe fluoride adsorption onto MACFBLcarbon material. Finally, fluoride desorption study and impact of various counter anions has been investigated for the removal of fluoride onto MACFBL material.

Table 1. Empirical kinetic equations and its parameters [31, 32].

Kinetic Models	**Kinetic equations**	**Equation Numbers**
Pseudo first-order kinetic model	$log(q_e - q_t) = \log q_e - \left(\frac{k_1}{2.303}\right) t$	(3)
Pseudo second-order kinetic model	$\left(\frac{t}{q_t}\right) = \left(\frac{1}{k_2 q_c{}^2}\right) + \left(\frac{1}{q_e}\right) t$	(4)
Intra-particle diffusion model	$q_t = k_d t^{1/2} + C$	(5)
Elovich kinetic model	$q_t = \left(\frac{1}{B}\right) ln\, AB + \left(\frac{1}{B}\right) ln\, t$	(6)

4. Result and Discussion

4.1. Characterizationof adsorbent material

Table 2. provides the physicochemical characterization results of the chemically activated microwave-assisted carbonized Ficus benghalensis leaf (MACFBL), i.e. the proximate and ultimate analysis. The Physico-chemical results obtained by characterization of MACFBL shows that the chemical activation and microwave treatment achieves the tremendous adsorbent carbon material required for fluoride adsorption.

4.2. Fourier transforms infrared (FTIR) study

Figure 2(a) displays the FTIR spectrum of the MACFBL carbon material before and after fluoride adsorption. The number of peaks reflects MACFBL's adsorptive existence The peaks in the region 3700 to 3400 cm^{-1} is because of the existence of –N-H and –O-H stretching vibrations. The peaks in the region 2900 to 2500 cm^{-1} represents the $-CH_2$ symmetrical and asymmetrical stretching. The peak region from 1700 to 1400 cm^{-1}reveals the appearance of the carbonyl group of ketones, esters and amide. The peaks due to –N-H deformation and bending were observed in the region 1400 to 1500 cm^{-1}. Around 1200 to 500 cm^{-1} position, peaks observed due to the existence of –C=S, -C-N, –C-H -C-O and -C-C- stretching vibrations and deformations. The shift in peak values and intensity band decrease might be because of the appearance of a chemical relationship between the functional groups which is in attendance on the surface of MACFBL. The shifts in the spectra demonstrate that MACFBL will be a helpful adsorbent material in the fluoride adsorption process. The IR band which either disappeared or decreases/increases in intensity after adsorption may engage in fluoride binding. Similar perceptions are accounted for by various scientists for fluoride adsorption [29, 30, 31, 32].

Table 2. Proximate and Ultimate analysis of MACFBL material

Proximate Analysis			Ultimate analysis		
Sr. No.	**Parameters**	**Values**	**Sr. No.**	**Parameters**	**Values**
1.	Bulk density (gm.cm^{-3})	0.39	1.	Carbon %	39.74
2.	Moisture content %	4.28	2.	Hydrogen %	1.09
3.	Ash content %	14.73	3.	Nitrogen %	0.55
4.	Volatile matter content %	32.65	4.	Sulphur %	00
5.	Fixed carbon content %	48.34	5.	Oxygen %	58.62
6.	pH	7.63	6.	Surface Area (m^2.g^{-1})	109.41
7.	Water Soluble Matter (%)	1.24	7.	Average Pore Diameter (A^O)	90.72
8.	Acid soluble matter (%)	3.73	8.	Total Pore Volume (cc .g^{-1})	0.248

4.3. X-ray diffraction (XRD) study
The XRD spectrum of MACFBL adsorbent before and after the removal of fluoride as presented in Figure 2(b). The strong main peak indicates the existence of a very much organized crystalline structure and noises in the XRD patterns show the amorphous nature of carbon material before and after fluoride adsorption. The strength of the well-structured peaks is also diminished to some extent. The XRD spectrum shows the fluoride adsorption was recognized by the physisorption process on the upper part of the MACFBL surface [32, 33].

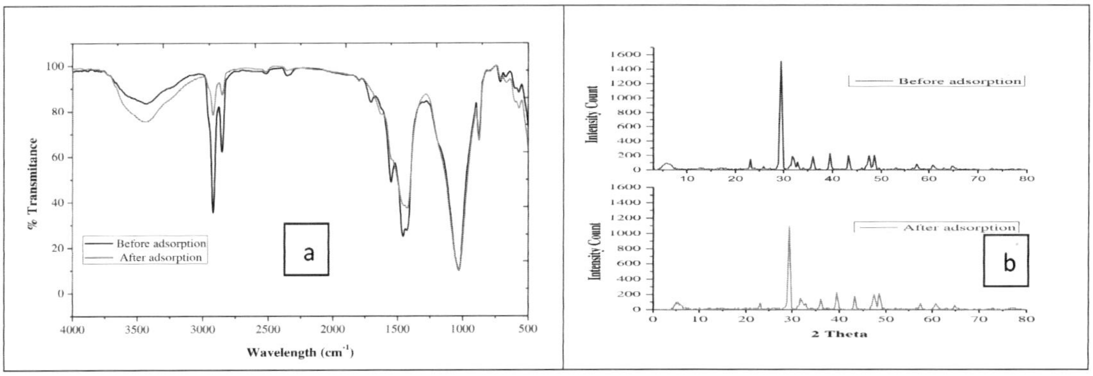

Figure 2. (a) Fourier-Transform-Infra-red (FTIR) spectra, (b) X-ray diffraction spectra of MACFBL before and after removal of the fluoride.

4.4. Scanning electron microscope (SEM) study
SEM micrographs of the MACFBL material before and after the adsorption of fluoride are seen in Figure 3 (a, b). SEM pictures of MACFBL shows that the adsorbent material has a rough surface with nearly non – compact structure. It is apparent that the adsorbent has a significant number of pore spaces, where proper conditions exist for fluoride ions to be caught and adsorbed into these pores. The bright spot demonstrates the existence of tiny holes on the crystalline structure of activated carbon material and after handling with fluoride the bright spots converts the black suggests the adsorption of the fluoride on the carbon surface through the physisorption process [30, 32, 33, 34].

4.5. Energy-Dispersive .X-ray (EDX) study

The EDX outcomes of the MACFBL material shown in Figures 3(a) and 3(b) indicate the variations between the number of elements in the adsorbent carbon material before and after adsorption of the fluoride. The fluoride content found in the EDX pattern for adsorbent material after contacting with adsorbate in aqueous solution indicates towards the adsorption of fluoride.

Figure 3.SEM-EDX monograph of MACFBL (a) before and, (b) after fluoride adsorption.

4.6. Effect of the contact time

The impact of the contact time on MACFBL was considered by varying the contact time from 30 min to 300 min for 2mg/l -10 mg/L fluoride concentrations by keeping other adsorption parameters constants (Figure 4). Adsorption of fluoride started at 30 min with 51.50 % and reached 86.50% at 150 min and the amount of fluoride adsorbed was 0.360 mg/g. No major improvement in the adsorption of fluoride was detected after 150 min. The underlying fast rate of adsorption of fluoride was might be due to the existence and of the availability of the extensive surface of the adsorbent material for fluoride particles present in the aqueous solution. The later moderate fluoride adsorption rate was might be because of the electrostatic obstruction due to already adsorbed fluoride ions and moderate pore diffusion of the particles [35,36]. So up to the contact time of 150 min the 86.50 % fluoride adsorption happens and after 150 min, the adsorption was expanded but the rate of adsorption was moderate as compared with the early rate of adsorption and the equilibrium point was occurred after 300 min. As a result, the minimum contact time for maximum fluoride reduction on the MACFBL adsorbent material was set at 150 minutes. A similar interpretation was achieved for the fluoride removal for initial concentrations of fluorides of 4mg/L, 6mg/L, 8 mg/L, and 10mg/L, but, there was a decline in per cent removal efficiency of fluoride.

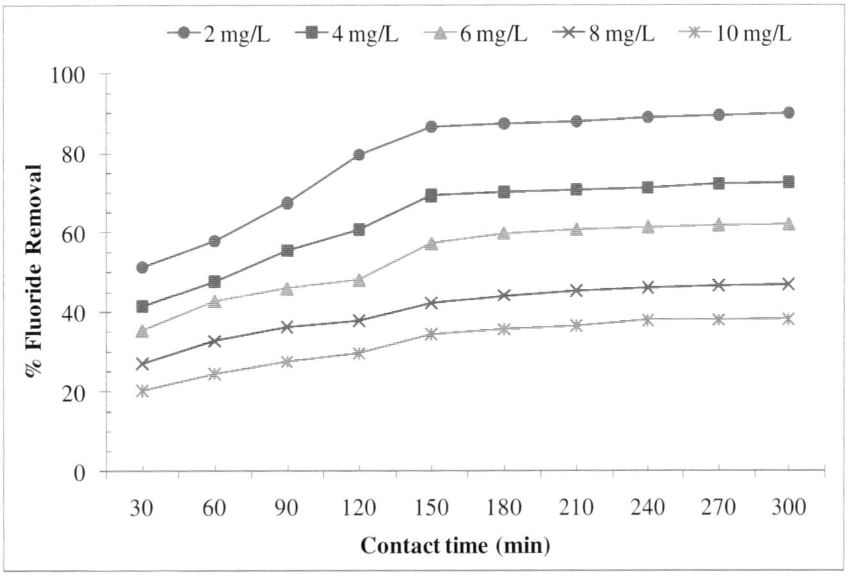

Figure 4. Effect of contact time on the fluoride adsorption by MACFBL.
(pH: 5, adsorbent dosage: 5 g/L, agitation speed: 120 strokes/min, initial fluoride concentration: 2 - 10 mg/L, temperature: 303 K)

4.7. Kinetics study for fluoride adsorption

The kinetic examination into the adsorption of fluoride from water plays an important role since it gives a significant understanding into the response pathways and mechanisms of the adsorption process of fluoride. With pseudo first-order, pseudo second-order, Intra-particle diffusion and Elovich models, the potential rate-restricting strides and probable kinetic mechanism of the adsorption of fluoride on MACFBL carbon material were investigated. Table 1 describes the observed kinetic equations and kinetic parameters of adsorption kinetic models for fluoride removal.

4.7.1. Pseudo first-order kinetic model for fluoride adsorption

Figure 5 presents a plot of log ($q_e - q_t$) against t. The plot shows a straight line for the first-order kinetics process, and Table 3 shows the outcomes of the fluoride adsorption on MACFBL. Using equation 3, the pseudo first - order rate constant (k_1) and the equilibrium adsorption power (q_e) were determined at different fluoride concentrations along with the correlation coefficients (R^2). The first-order rate constant (k_1) values are extending from 0.520 to 0.620 min^{-1} over the fluoride concentration range from 2 mg/L to 10 mg/L. A wide dissimilarity involving between the theoretical values and experimental values of equilibrium fluoride adsorption capacity, q_e, is seen from the kinetic results (Table 3) in the fluoride concentration range (2 mg / L-10 mg / L), suggesting weak applicability of the pseudo first - order kinetic model to the experimental results.

4.7.2. Pseudo second-order kinetics model for fluoride adsorption

The pseudo second-order kinetic model parameters of fluoride adsorption were estimated by utilizing empirical equation 4. The linear plot of t/q_t against t at various fluoride concentrations (from 2 mg/L to 10 mg/L) is made known in Figure 6. The second-order rate constant (k_1) for fluoride adsorption, and the equilibrium fluoride adsorption capacity (q_e) are calculated from the plot of t/qt against t. Table 3 describes the respective approximate parameters of the second order kinetic model. It is evident from the kinetic results that, relative to the pseudo-first - order kinetic model, the pseudo-second - order kinetic model displayed excellent linearity with admirable correlation coefficient values (R^2> 0.992) at the fluoride concentration tested. With only minor variations, the q_e values calculated from the second-order kinetic plots are incredibly similar to the experimental values of q_e. This demonstrates that the mechanism of fluoride adsorption fits the kinetic pseudo second - order relation. The pseudo second-order fluoride adsorption rate constant, k_2, values ranging in between 0.322 $gmg^{-1}min^{-1}$ to 0.899 $gmg^{-1}min^{-1}$.

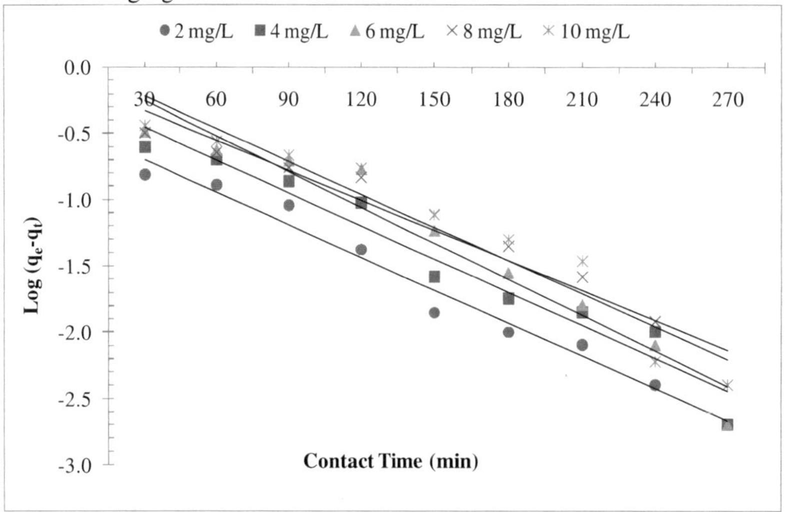

Figure 5. Linear Plots of the Pseudo first-order kinetic for fluoride adsorption on to MACFBL.

4.7.3. Intra-particle diffusion model for fluoride adsorption

The relationship based on the elimination of contaminants from aqueous solution was examined using the Weber-Morris intra-particle diffusion model. The linear plot results (Figure 7) obtained by the use of equation 5 and Table 3 refer to the external take-up of the surface. It may suggest that intra-particle diffusion was involved in the adsorption of fluoride on the surface of MACFBL. Based on these findings, but it was not the sole rate-determining step and that during fluoride adsorption, various other system processes often took part. The kinetic plot in the present case shows reasonably high rate coefficients and also better linearity. The comparative pattern has been reported for fluoride adsorption of fluoride by researchers [36, 37, 38].

4.7.4. Elovich kinetic model for fluoride adsorption

In order to validate the suitability of the Elovich kinetic equation (Figure 8 and Table 3), the adsorption of the fluoride from the aqueous solution into the MACFBL carbon material, a linear plot of q_t versus Log t at various contact time for the removal of the fluoride is examined. From the intercept and slope of the linear plot, the parameters of the Elovich kinetic constants A and B were estimated. The correlation coefficient values (R^2 = 0.937-0.984) over the initial fluoride concentration range for the Elovich equation indicate that fluoride diffusion fits the kinetic process proposed by the Elovich model and the rate-deciding step is the diffusion in nature [39]. In fluoride adsorption to the surface of MACFBL adsorbent material, this diffusional rate-limiting mechanism is more clearly identifiable.

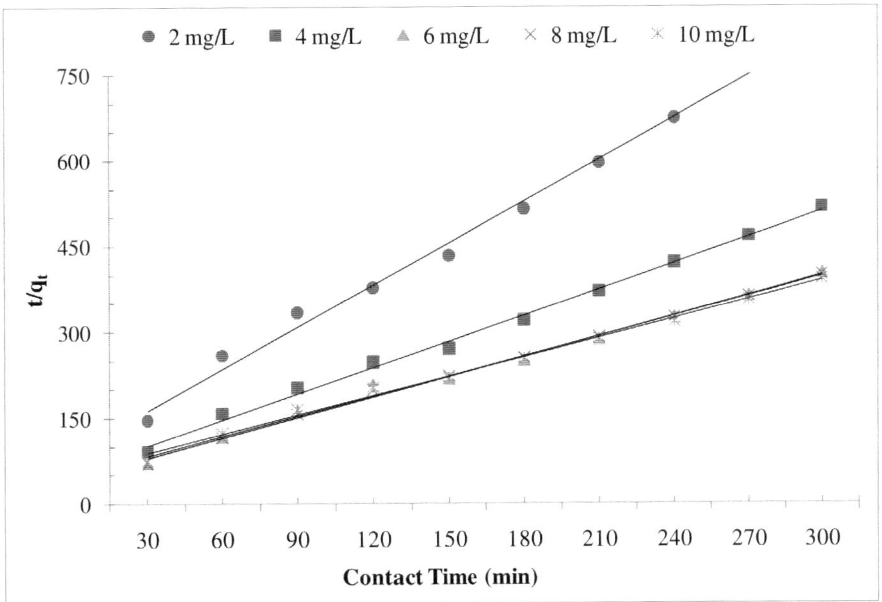

Figure 6. Linear plots of the Pseudo second-order kinetic for fluoride adsorption on to MACFBL.

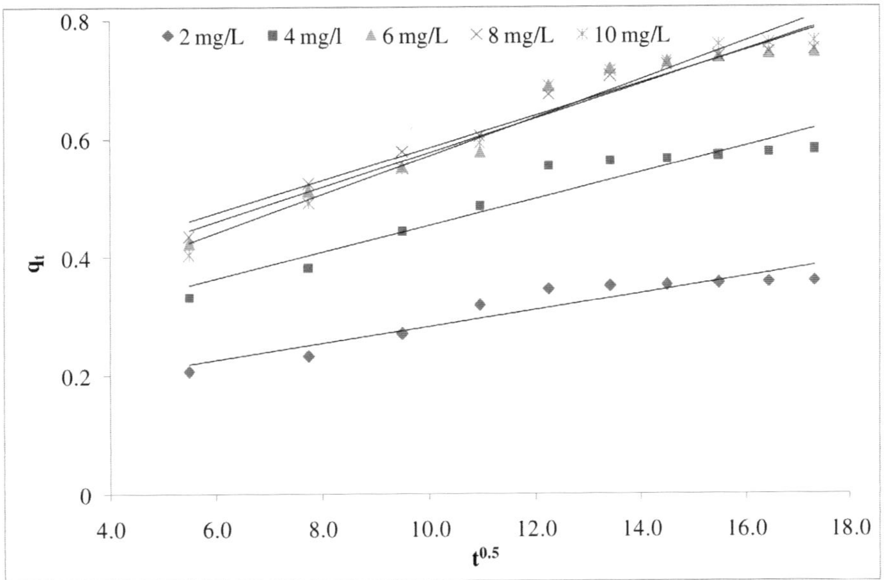

Figure 7. Linear plots of the Intra-particle diffusion model for fluoride adsorption on to MACFBL.

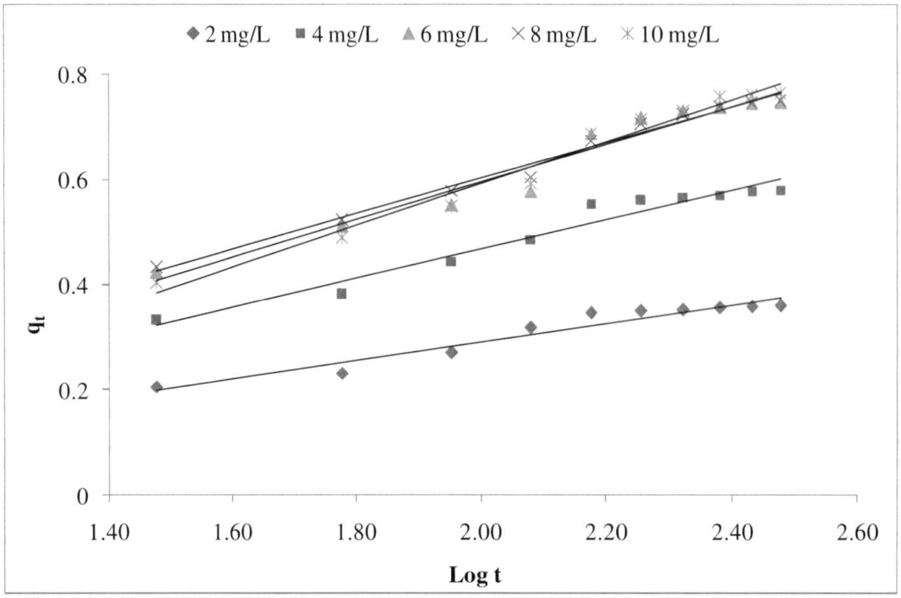

Figure 8. Linear Plots of the Elovich kinetic model for the fluoride adsorption on to MACFBL.

4.8. Effect of temperature on fluoride adsorption

The analysis of the impact of temperature variation is significant in order to find out the viability of the fluoride adsorption process.The impact of temperature on the evacuation of the fluoride using MACFBL material was examined by performing the adsorption experiments at four different temperatures 303, 313, 323 and 333 K, by keeping other parameters constant for initial concentration of the fluoride from 2 mg/L to 10 mg/L. The fluoride adsorption potential was found to decrease from 0.360 mg/g to 0.272 mg/g with an initial fluoride concentration of 2 mg/L as the temperature increased from 303 K to 333 K. (Table 4 of Figure 9). Comparative patterns were found for higher fluoride concentrations. This indicates that the mechanism of the fluoride adsorption is an exothermic process [40]. It is seen in Figure 3.21 that low temperatures favour the elimination of fluoride. This may be due to the tendency for the fluoride ions to break apart with an expansion of the aqueous solution temperature from the solid phase to the bulk phase [36, 40].

4.9. Thermodynamic study

Thermodynamic parameters such as free energy change (ΔG), enthalpy changes (ΔH) and entropy change (ΔS) can be calculated by using equations (7), (8), (9) for the thermodynamic analysis to learn the feasibility of the adsorption of fluoride. In order to discover the values of ΔS and ΔH from the slope and intercept (Figure 10), a linear plot of Log K versus 1/T can be applied. Table 4 displays the Gibbs free energy (ΔG) for the adsorption of fluoride on MACFBL carbon content at all temperatures. The outcomes of the thermodynamics investigation describe the estimated values of ΔH, ΔG, and ΔS. The negative values of enthalpy change have shown the exothermic and random existence of the mode of adsorption[41], and the negative values of entropy change have shown that the process of adsorption is enthalpy-driven[41]. For the lower temperatures and lower fluoride concentrations, the ΔG values observed in the present sample were negative. The change in free energy (ΔG) values was positive as the temperature expanded and the adsorption turned out to be less ideal. The present research showed that at a lower temperature, the fluoride adsorption mechanism was more favourable and feasible. Comparable negative and positive estimates of ΔG for adsorption of fluoride on various adsorbents have been reported by researchers [41, 42].

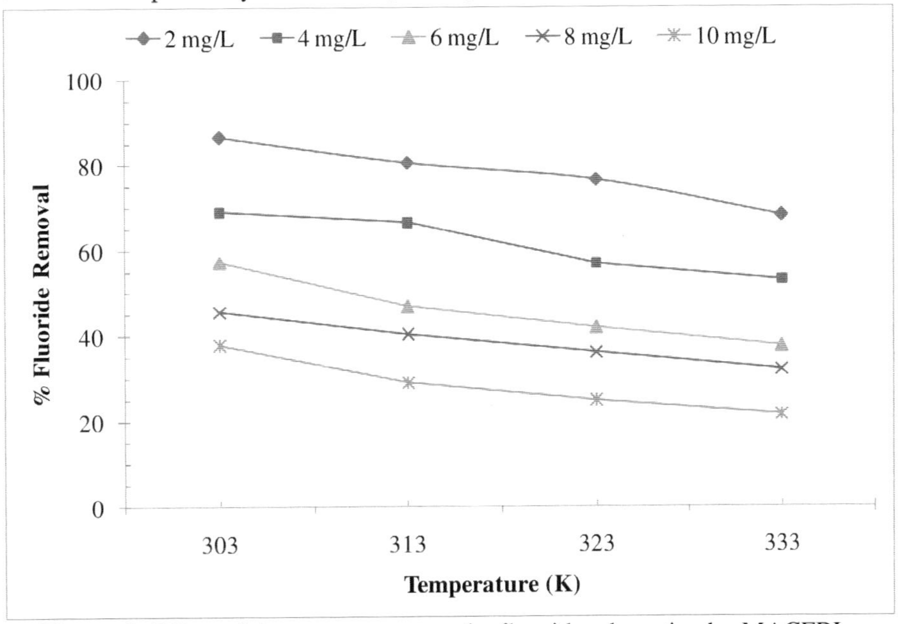

Figure 9. Effect of the temperature on the fluoride adsorption by MACFBL.
(pH: 5, adsorbent doses: 5 g/L, agitation speed; 120 strokes/min, contact time : 150 min, fluoride concentrations : 2 – 10 mg/L)

Table 3. Kinetic parameters of fluoride adsorption on MACFBL

Kinetic model	Kinetic parameters	Initial concentrations of fluoride (mg/L)				
		2	4	6	8	10
Pseudo-first-order	q_e (mg/g) [Plot]	0.352	0.621	1.035	0.787	1.064
	q_e (mg/g) [Expt.]	0.360	0.580	0.746	0.750	0.764
	k_1 (min^{-1})	0.567	0.573	0.620	0.520	0.571
	R^2	0.978	0.953	0.943	0.947	0.925
Pseudo-second-order	q_e (mg/g) [Plot]	0.354	0591	0.735	0.758	0.761
	q_e (mg/g) [Expt.]	0.360	0.580	0.746	0.750	0.764
	k_2 (g/ mgmin)	0.899	0.515	0.377	0.397	0.322
	h (mg/ g min) (10^{-1})	0.113	0.180	0.210	0.228	0.187
	R^2	0.994	0.995	0.992	0.997	0.994
Intra-particle diffusion	k_d (mg/g $min^{0.5}$)	0.014	0.220	0.029	0.027	0.032
	C	0.142	0.231	0.284	0.310	0.247
	R^2	0.887	0.912	0.935	0.917	0.958
Elovich	A(mg/g min (10^{-1})	0.349	0.583	0.718	0.892	0.534
	B (g/mg)	13.160	8.314	6.433	6.813	5.801
	R^2	0.937	0.954	0.952	0.984	0.973

$$\Delta G = -R\ T\ \ln K \tag{7}$$

$$\Delta G = \Delta H + T\ \Delta S \tag{8}$$

$$\ln K = \frac{\Delta S}{R} - \frac{\Delta H}{R\,T} \tag{9}$$

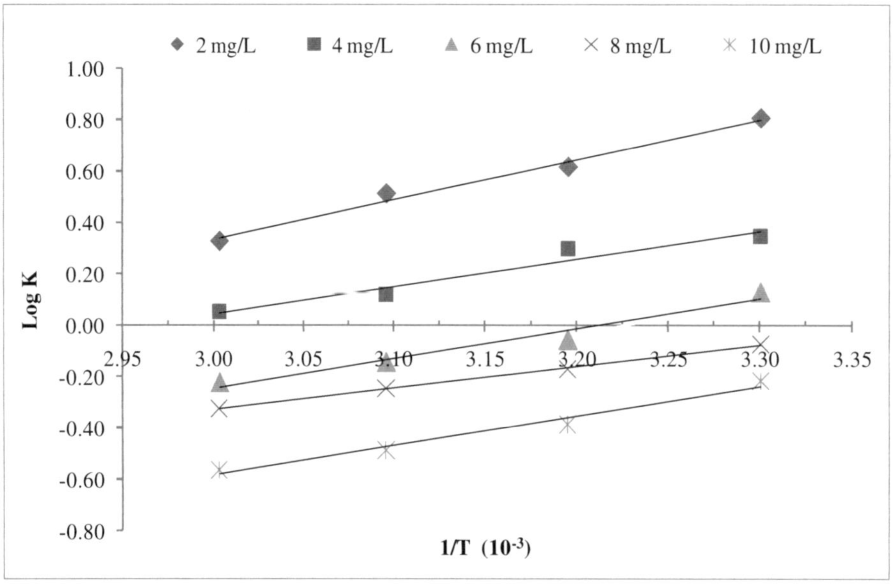

Figure 10. Linear plots of Log K against 1/T for fluoride adsorption by MACFBL.

4.10. Effect of counter anions on fluoride removal

Removal of fluoride was performed experimentally in the presence of typical counter ions, such as sulphate, bicarbonate, chloride, nitrate, and phosphate, which are normally found in water. The counter ion concentration ranged from 0 to 500 mg/L at an initial dosage of 2 mg/L at a set pH 5 and a contact time of 150 min at a dosage of 5 g / L, retaining an agitation speed of 120 strokes / min at a temperature of 303 K. Figure 11 showed that there was a significant effect on the removal of fluoride in the presence of these counter ions. The presence of bicarbonate and phosphate ions, however, resulted in a drop in the fluoride removal percentage. This may be a direct result of the opposition of these counter ions with fluoride for adsorption sites [43, 44]. The unique selective nature of the adsorbent material depends on the various properties of the counter ions such as polarizability, size, electronegativity difference charge of ions, etc. [45, 46]. The descending hindrance observed for different counter ions for removal of fluoride onto MACFBL material is in the following order, Bicarbonate >Phosphate >Nitrate>Sulphate >Chloride.

Table 4. Thermodynamic properties for adsorption of fluoride onto MACFBL

C_o (mgL^{-1})	ΔH (k J $mole^{-1}$)	ΔS (kJ mol^{-1} K^{-1})	ΔG (kJ $mole^{-1}$)			
			303 K	313K	323 K	333 K
2	-29.754	-0.082	-4.680	-3.690	-3.170	-2.087
4	-20.544	-0.061	-2.016	-1.784	-0.729	-0.332
6	-22.095	-0.071	-0.744	0.330	0.885	1.414
8	-16.007	-0.054	0.429	1.028	1.530	2.071
10	-21.961	-0.077	1.265	2.317	3.008	3.586

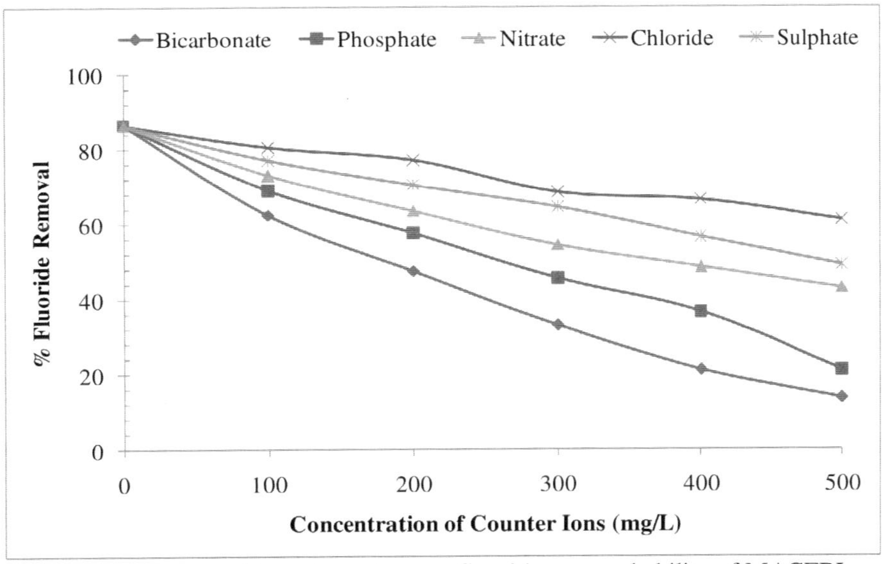

Figure 11. Impact of Counter anions on fluoride removal ability of MACFBL
(pH: 5, adsorbent doses: 5 g/L, agitation speed: 120 strokes/min, contact time: 150 min, temperature: 303 K, fluoride concentrations: 2 – 10 mg/L)

4.11. Desorption Study

A desorption or regeneration study is also essential since it is valuable in the recycling of the MACFBL material and recovery of the fluoride. Desorption of fluoride utilizing with 0.1 & 0.2 mol/L of NaOH, KOH and Na_2CO_3 alkali solutions was considered for the investigation. The desorption arrangement was operated at the optimum condition based on the batch experiment results. 50 ml test solution was taken from the effluent during desorption study and estimate for fluoride by using an ion-selective method. Figure12, demonstrates the increase in desorption from 0.1M to 0.2 M alkali solutions. Results indicate that 0.2 mol/L alkali solutions regenerate fluoride-rich carbon up to 69.5 – 62.5 %. It was found that the MACFBL carbon material has been effectively regenerated with alkali solutions [47-50].

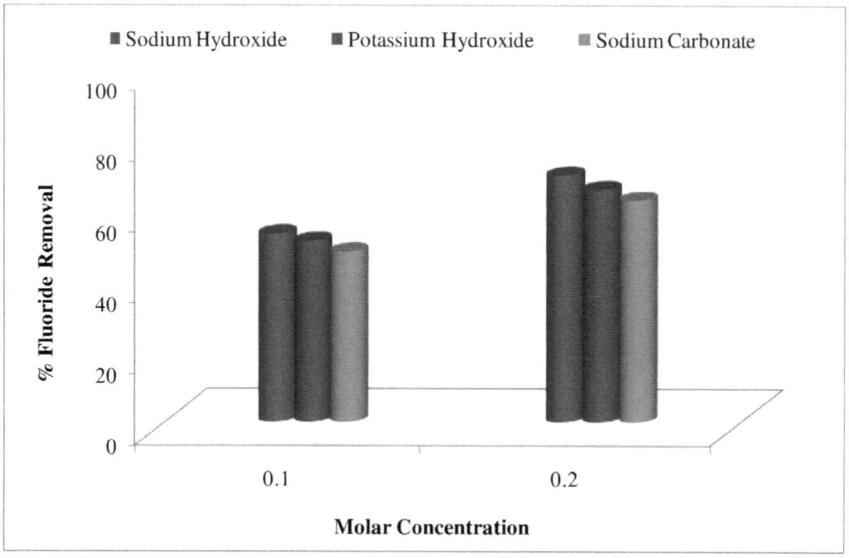

Figure 12. Desorption ability of MACFBL using alkali solutions.
(pH: 5, adsorbent doses: 5 g/L, agitation speed: 120 strokes/min, contact time; 150 min, temperature: 303 K, fluoride concentration: 2 mg/L)

5. Conclusion

In the current investigation, a new chemically impregnated carbon material from leaves of *Ficus benghalensis* (MACFBL) has been developed for expulsion of fluoride from synthetic solution of fluoride. The adsorbent material could be simply prepared from leaves *Ficus benghalensis*that are locally available everywhere. It has developed as an efficient, porous and economically reasonable carbon material which demonstrated the highest percentage removal of fluoride i.e. 86.5 %. The developed microwave treated and chemically activated adsorbent material (MACFBL) was characterized to investigate properties of adsorbent material by proximate analysis like bulk density, volatile matter, ash content, moisture content, water and acid soluble matter, and fixed carbon; ultimate and instrumental analysis like C, H, N, S element analysis, BET surface analysis, SEM, FTIR, EDX and XRD study. The characterization results show tremendous adsorbent properties for MACFBL carbon material. The fluoride adsorption rate was additionally examined with pseudo first-order model, pseudo second-order, Intra-particle diffusion, and Elovich kinetic models. The experimental finding was found to suit superior to pseudo second-order over other kinetic models. Thermodynamic investigation shows that fluoride adsorption favours at the lower temperature and the adsorption of fluoride was spontaneous, feasible and exothermic. Finally, it was concluded that MACFBL carbon material has adequate potential to extract fluoride from water and also 69.50 % fluoride can be recovered from the adsorbent surface by using a 0.2 M NaOH solution.

Acknowledgment

The author is grateful to the Principal, Dr Ambedkar College, Nagpur for providing laboratory facilities for the conduct of this research work.

References

[1] Annadurai ST,Arivalagan P, Sundaram R, Mariappan R, MunusamyA P 2019 Batch and column approach on biosorption of fluoride from aqueous medium using live, dead and various pretreated Aspergillus niger (FS18), biomass, Surf and Interf **15** 60–69.

[2] Liang S, Xue Y, Gao B and Yang K 2017 Removal of fluoride from aqueous solution by TiO_2-based composites, J. Taiwan Inst. Chem. Eng **74** 205–210.

[3] Tomar V, Prasad S, Kumar D 2014 Adsorptive removal of fluoride from aqueous media using Citrus limonum (lemon) leaf, *Microchemical Journal***112** 97-103.

[4] Li YH, Wang S, Cao A, Zhao D, Zhang X, Xu C 2001 Adsorption of fluoride from water by amorphous alumina supported on carbon nanotubes *Chem Phys Lett***350** 412–416.

[5] Shahid MK, Kim JY, Choi Y G 2019 Synthesis of bone char from cattle bones and its application for fluoride removal from the contaminated water *Groundwater for Sustainable Development***8** 324-331.

[6] World Health Organization (WHO), Guidelines for Drinking Water Quality, fourth ed., (2011).

[7] Bureau of Indian Standards (BIS), Indian standard: drinking water specification, IS: 10500, New Delhi (2003).

[8] Nigri EM, Bhatnagar A, Rocha S D F 2017 Thermal regeneration process of bone char used in the fluoride removal from aqueous solution *J Clean Prod***142** 3558–3570.

[9] Jagtap S, Yenkie M K, Labhsetwar N, Rayalu S 2012 Fluoride in drinking water and defluoridation of water *Chem Rev***112** 2454–2466.

[10] Araga R, Soni S, Sharma C S 2017 Fluoride Adsorption from Aqueous Solution using Activated Carbon Obtained from KOH-treated Jamun (Syzygium cumini) Seed *J Environ Chem Eng***5** 5608-5616.

[11] MohantaD, Ahmaruzzaman Md 2018 Bio-inspired adsorption of arsenite and fluoride from aqueous solutions using activated carbon@SnO2 nanocomposites: Isotherms, kinetics, thermodynamics, cost estimation and regeneration studies, *J Enviro. Che. Eng***6** 356–366.

[12] Cai H-m, Chen G-j, Peng C-y, Zhang Z-z, Dong Y-y, Shang G-z, Zhu X-h, Gao H-j, Wan X-C 2015 Removal of fluoride from drinking water using tea waste loaded with Al/Fe oxides: a novel, safe and efficient biosorbent *Applied Surface Science***328** 34-44.

[13] Ahamad KU, Singh R, Baruah I, Choudhury H, Sharma M R 2018 Equilibrium and Kinetics Modeling of Fluoride Adsorption onto Activated Alumina, Alum and Brick Powder, *Groundwater for Sustainable Development*7452-458.

[14] Telkapalliwar NG, Shivankar V M 2016 Removal of fluoride from aqueous solution by using inexpensive bark and wood based bioadsorbents : A comprehensive review, IJARIIE **2** 3447-3457.

[15] Telkapalliwar N G, Shivankar V M 2016 Review on adsorption of fluoride from aqueous solution by using low cost leaf based bioadsorbents*Int. J. of Modern Trends in Eng and Res***03**40-52.

[16] Telkapalliwar N G, ShivankarVM 2016 Fruit Peel and Shell as a Low-cost Bioadsorbents for Fluoride Removal from aqueous Solution: A Review, *Inter J Innov. Res in Sci Eng.and Technol* **5** 10248-10257.

[17] Telkapalliwar N G, Shivankar V M 2016 Growing Approach in Adsorption of Fluoride from Aqueous Solution by using Inexpensive Seed Based Biomass: Review *J Chem Bio and Phy Sci***6** 946-957.

[18] Ammavasi N, Kumar SK, Mariappan R 2017 Investigation of lanthanum impregnated cellulose, derived from biomass, as an adsorbent for the removal of fluoride from drinking water, *Carbohydrate Polymers***176** 402-410.

[19] Murambasvina G, Mahamadi C 2020 Effective fluoride adsorption using water hyacinth beads doped with hydrous oxides of aluminium and iron, *Groundwater for Sustainable Development* **10** 100302.

[20] Mondal N K, BhaumikR, DattaJK 2015 Removal of fluoride by aluminum impregnated coconut fiber from synthetic fluoride solution and natural water, *Alexandria Eng. J*54 1273-1284.

[21] MeiL, Qiao H, KeF, Peng C, HouR, Wan X, Cai H 2019 One-step synthesis of zirconium dioxide-biochar derived from Camellia oleifera seed shell with enhanced removal capacity for fluoride from water *Applied Surface Science* 144685.

[22] Khound N J,Bharali R K 2018 Biosorption of Fluoride from Aqueous Medium by Indian Sandalwood (Santalum Album) Leaf Powder, *J Environ Chem Eng***6** 1726-1735.

[23] Ghosh S B, Mondal N K 2019 Application of Taguchi method for optimizing the process parameters for the removal of fluoride by Al-impregnated Eucalyptus bark ash *Environmental Nanotechnology, Monitoring and amp; Management***11**100206.

[24] Gaikwad MS, Balomajumder C 2017 Tea waste biomass activated carbon electrode for simultaneous removal of Cr(VI) and fluoride by capacitive deionization, *Chemosphere*,**184** 1141-1149.

[25] Chen C-L, Park S-W, Su J F, Yu Y-H, Heo J-E, Kim K-D, Huang CP 2019 The adsorption characteristics of fluoride on commercial activated carbon treated with quaternary ammonium salts (Quats), *Science of the Total Environment***693** 133605.

[26] Bharali RK, Bhattacharyya KG 2015 Biosorption of fluoride on Neem (Azadirachta indica) leaf powder, *J. Environ. Chem. Eng***3** 662-669.

[27] George AM, Tembhurkar A R 2019 Analysis of equilibrium, kinetic, and thermodynamic parameters for biosorption of fluoride from water onto coconut (Cocos nucifera Linn.) root developed adsorbent*Chinese Journal of Chemical Engineering***27** 92–99.

[28] Edmunds W M, Smedley P L 1996Environmental Geochemistry and Health *Geological Society Special Publication***113** 91-105.

[29] USEPA, R.E.D. FACTS, Cryolite, EPA-738-F-96-016, United States Environmental Protection Agency, 1996.

[30] Reeves TG, Water Fluoridation: A Manual for Engineers and Technicians, United States Department of Health and Human Services, Centres for Disease Control and Prevention, 1986.

[31] Telkapalliwar NG, Shivankar VM 2018Adsorption of Zinc onto Microwave assisted carbonized Acacia nilotica bark *Material Today Proceedings***5** 22694-22705.
[32] Telkapalliwar NG, Shivankar VM 2019 Data of characterization and adsorption of fluoride from aqueous solution by using modified Azadirachta indica bark, *Data in Brief***26** 104509. https://doi.org/10.1016/j.dib.2019.104509
[33] Karthikeyan G, Ilango SS 2007 Fluoride sorption using Morringa indica-based activated carbon, *Iranian J Environ Health Sci& Eng***4** 21-28.
[34] Al-Ghouti M A, Khraisheh MAM, Allen SJ, Ahmad MN 2003 The Removal of Dyes from Textile Wastewater: A Study of the Physical Characteristics and Adsorption Mechanisms of Diatomaceous Earth *J Environmental Management* **69** 229–238.
[35] Kumar S, Gupta A, Yadav J P 2008 Removal of fluoride by thermally activated carbon prepared from neem (Azadirachta indica) and kikar (Acacia arabica) leaves, *J Environ Biology***29** 227-232.
[36] Sujana M G, Anand S 2011 Fluoride removal studies from contaminated ground water by using bauxite *Desalination***267** 222-227.
[37] Aharoni C, Sparks DL, Levinson S, Ravina I 1991 Kinetics of soil chemical reactions: Relationship between empirical equations and diffusion models, *Soil Sci Soc Am J***55** 1307-1312.
[38] Rudzinksi W, Panczyk P,Schwarz JA, C. I.Contescu (Eds.), Surfaces of Nanoparticles and Porous Materials, Dekker, New York, (1998) 355-391.
[39] AlagumuthuG, Veeraputhiran V, Venkataraman R 2011 Fluoride sorption using Cynodon dactylon-based activated carbon. *Hem Ind***65** 23–35.
[40] Murugan M, Subramanian E 2006Studies on defluoridation of water by Tamarind seed, an unconventional biosorbent,*J Water and Health***4** 453-461.
[41] SahaP, Chowdhury S, Gupta S, Kumar I 2010 Insight into adsorption equilibrium, kinetics and thermodynamics of Malachite Green onto clayey soil of Indian origin*Chem. Eng. J***165**874-882.
[42] Xu X, Li Q, Cui H, Pang J, Sun L, An H, Zhai J 2011 Adsorption of fluoride from aqueous solution on magnesia-loaded fly ash cenospheres *Desalination***272** 233–239.
[43] Maliyekkal SM, Shukla S, Philip L, Nambi I M 2008Enhanced fluoride removal from drinking water by magnesia-amended activated alumina granules *Chem Eng J***140** 183-192.
[44] Kamble SP, Jagtap S, Labhsetwar NK, ThakareD, Godfrey S, DevottaS, Rayalu SS 2007 Defluoridation of drinking water using chitin, chitosan and lanthanum-modified chitosan, *Chem Eng J* **129** 173-180.
[45] Suneetha M, Bethanabhatla SS, Ravindhranath K 2015Removal of fluoride from polluted waters using active carbon derived from barks of Vitex negundo plant *J Anal Sci and Technol***6**1-19.
[46] Alagumuthu G, Rajan M 2010 Equilibrium and Kinetics of Adsorption of Fluoride onto Zirconium Impregnated Cashew Nut Shell Carbon. *Chem Eng J* **158** 451-457.
[47] Onyango M S, Kojima Y, Aoyi O, Bernardo E C, Matsuda H 2004 Adsorption equilibrium modeling and solution chemistry dependence of fluoride removal from water by trivalent-cation-exchanged zeolite F-9, *J Colloid Interface Sci***279** 341-350.
[48] Gómez-Hortigüela L, Pérez-Pariente J, García R, Chebude Y, Díaz I 2013Natural zeolites from Ethiopia for elimination of fluoride from drinking water*Sep Purif Technol***120** 224-229.
[49] Chang CF, Chang CY, Hsu TL 2011 Removal of Fluoride from Aqueous Solution with the Superparamagnetic Zirconia Material *Desalination***279** 375-382.
[50] Kagne S, Jagtap S, Dhawade P, Kamble S P, Devotta S, Rayalu S S 2008 Hydrated cement: a promising adsorbent for the removal of fluoride from aqueous solution, *J Hazard Mater***154** 88-95.

Electrical and Electrochemical studies of Polyurethane diol/ Polycaprolactone- Iron Oxide nanocomposites

A Suhasini[1], K P Vinod Kumar[2], A Maria Sheela[1] and N SheenKumar[3]

[1]Assistant Professor, Department of Chemistry, St.Xavier's Catholic College of Engineering, Nagercoil,Tamilnadu,India
[2] Assistant Professor, Department of Chemistry, University College of Engineering,Nagercoil,Tamilnadu,India
[3] Assistant Professor, Department of Physics, St.Xavier's Catholic College of Engineering, Nagercoil,Tamilnadu,India

Corresponding email: suhasiniaugustin@gmail.com

Abstract: Electrically conducting composite materials consisting of different weight percentage of iron oxide nanoparticles dispersed in Polyurethane diol/Polycaprolactone were prepared. As a function of temperature between 30-70°C and frequency in the range of 1 kHz to 1000 kHz the dielectric constant and conductivity of polymer nanocomposites were studied. The dielectric constant value increased with increase in temperature and decreased with increase in frequency. Conductivity was found to be in the order of 10^{-6} to 10^{-5} S/cm and increased with increase in temperature. Cyclic voltammetry investigations revealed that the material has the potential to perform as an alternative catalyst for fuel cell applications.

1. Introduction

Polyurethanes (PU) are multipurpose materials because they can be easily shaped and established excellent workability. Therefore, they are broadly used in various fields of applications like construction, electrical, automotive industry, electronics, footwear production and biomedical applications like heart valves, dialysis membranes [1], artificial joint, artificial blood vessel, fracture fixation, controlled release devices etc., [2,3]. Polyurethane diol (PUD) is the family of polyurethane and are used as a transformer for auto coatings, suspensions in wood, water-soluble resins and appliances [4]. Polycaprolactone (PCL) is one among the manmade biodegradable polymers used to make compostable polymeric electronic devices and in medical field as artificial bone, artificial skin, knee dislocation and containers for continuous drug delivery [5, 6]. Polymer blending is an important process to synthesize new materials, which have better properties than a single polymer. Polyurethane/ polycaprolactone blends are used in orthopaedic binding or molding, prosthetic socket cone production, radiotherapy patient immobilization and plastic reconstructive surgery [7, 8]. Polymer blending can combine organic and inorganic components to get different materials with desirable properties [9,10]. Polymer blends revealed greater conductivity and additional mechanical strength when compared to that of pure polymers [11-13]. The iron oxide nanoparticles have superior catalytic activity, super paramagnetic behaviour and large surface area, so they can be used in solar energy transformation, diagnostic changing, magnetic liquids, during delivery, biological separation, magnetic storage media, electrode materials, [14,27], electronics, purification processes and catalysis. The present work explains the effect of electrical and electrochemical properties of polymeric nanocomposites prepared with Fe_2O_3 nanoparticle incorporated in PUD/PCL blend. The nanocomposites prepared were investigated for mechanical, thermal, dynamic mechanical thermal analysis, magnetic, X-ray diffraction, HR-SEM and TEM and already reported [15].

2. Materials and methods

2.1 Materials

Polycaprolactone , average molecular weight ~ 45,000 was used after dehydration under vacuum at 60 °C for 4hrs. Polyurethane diol (PUD), average molecular weight ~320 was used after removal of water at 100 °C for 12 hours in a vacuum air oven. 4, 4'-diaminodiphenylmethane (DDM) and a catalyst Dibutyltindilaurate (DBTDL), Iron oxide nanoparticles were purchased from Sigma Aldrich and hexamethylene diisocyanate (HDI) from Fluka were used as received with no extra purification.

2.2 Methods

The nanocomposites prepared were fabricated in the form of pellet and enclosed by graphite to get a good conducting surface layer. The pellet size was measured by means of a travelling microscope. The dielectric constant and conductivity measurements were analyzed by means of an LCR meter (Agilent 4284 A). The CV measurements were approved by electrochemical workstation 604D, CHI Instruments, USA electrochemical analyzer.

2.3 Synthesis of PUD/PCL-Fe_2O_3 nanocomposites

40g of polyurethane diol was taken in a warmed mixer and heated at around 90°C and then 10 g of polycaprolactone was added. The mixture was heated at 120° C and agitated at a rotor speed of 415 rpm for 30 minutes constantly. Polycaprolactone pellet was completely miscible with the melted polyurethane diol. To this homogeneous mixture, 0.5g of Fe_2O_3 nanomaterial was added slowly and agitated at 415 rpm for 45 minutes. One hour of sonication allowed the nanoparticles to dissolve regularly throughout the blend. Heating was stopped and then 12 ml hexamethylene diisocyanate and 4g diaminodiphenylmethane were added and agitated. The whole content was relocated to the mold and dehydrated overnight at room temperature in a vacuum air oven and at 100 °C for 1hour.

2.4 Electrode preparation for electrochemical characterization

Electrochemical workstation 604D, CHI Instruments, USA electrochemical analyser with a three-electrode cell assembly consisting of a saturated calomel electrode as reference electrode, Pt wire served as supporting electrode and PUD/PCL-Fe_2O_3 nanocomposite slurry coated copper foil as working electrode with 1M KOH as the electrolytic solution were used for cyclic voltammetry measurements. 0.1mm thickness copper foil electrode was prepared with a dimension of 1 cm^2. 0.1g of the nanocomposite was dissolved in DMF and a colloidal solution was prepared by sonication for 15 min. Then the slurry was spread on the exterior of a copper foil and dried at room temperature to get a film on the surface of the copper foil.

3. Results and discussion

3.1 Temperature dependence dielectric constant

The relative permittivity variation of the polymer nanocomposites as a function of temperature over frequency of 1kHz,10kHz,100kHz and 1000kHz are depicted in figure 1(A-D). The dielectric performance with temperature variation is different for different temperature. It is clear that from the graphs, the dielectric constants are low at low temperature and when the temperature is raised, the value rises correspondingly. At room temperature, the effect of grain boundaries is dominant, which is reflected in the low magnitude of dielectric properties. Nevertheless, as the temperature is increased, the sizes of grains become more, which results in the increase of dielectric properties [16]. Furthermore, the relative permittivity (ε_r) of the pure blend was very low, when compared to various Fe_2O_3 nanoparticles loaded PUD/PCL blend. This is attributed to the fact that the dielectric properties of polymeric material are determined by the charge dispersal along with the statistical thermal motion of their polar groups [17]. In the case of polar polymers, the dielectric constant begins to drop at a certain temperature invariably for all frequencies, which is also observed in the nanocomposites prepared [18]. This can be attributed to the phase changes that occur around 50°C for different composites prepared at all frequency ranges. These variations in dielectric values can be attributed to the changes in orientation polarization caused by temperature change [19].

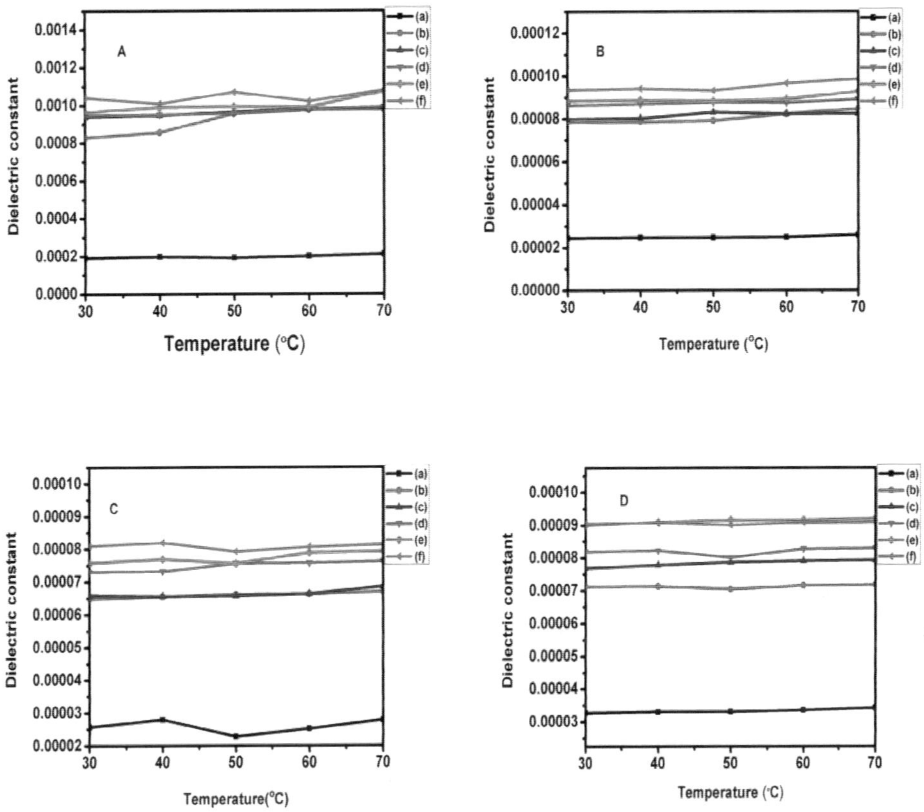

Figure 1. (A-D) relative permittivity and temperature dependence curves for the frequencies 1KHz, 10KHz, 100KHz, 1000KHz for nano iron oxide composites (a)Pure blend (b) 0.5 wt% nano iron oxide composite (c) 1.0 wt% nano iron oxide composite (d) 1.5 wt% nano iron oxide composite (e) 2.0 wt% nano iron oxide composite(f) 2.5 wt% nano iron oxide composite.

3.2 Frequency dependence dielectric constant

The dielectric constant variation with frequencies at 30°C and 70°C for dissimilar concentration of Fe_2O_3 nanoparticles are given in the figure 2 (A& B). It is observed from the figure, as the frequency increases the dielectric constant decreases. This change is due to the charge carriers located at the boundaries of the nanoparticles and polymer matrix forming space charges. Along the applied field, these space charges will orient themselves, while the electric field is applied. When the frequency of the applied field increases, polarization decreases and hence the dielectric constant decrease [17]. The dielectric constant value for pure blend at 30°C has extremely low value, but the addition of nano iron oxide filled blend shows approximately 10 times higher values. The low value for blend may be owing to the space charge division at sample –electrode border, whereas high values may be connected to the orientation polarization effects due to relaxing dipoles. The tiny increase at high temperature may be due to faster ion migration at high temperature above (T_m = 50°C) due to the polymer matrix phase change [20].

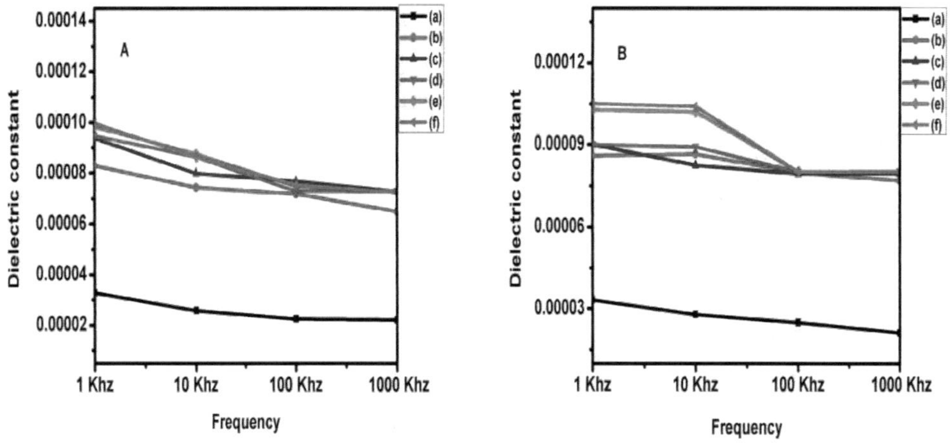

Figure 2. (A&B) Dielectric constant variation with frequency at(30°C) & 70° C for nano iron oxide composites (a) Blend (b) 0.5 wt% nano iron oxide composite (c) 1.0 wt% nano iron oxide composite (d) 1.5 wt% nano iron oxide composite (e) 2.0 wt% nano iron oxide composite (f) 2.5 wt% nano iron oxide composite.

3.3 Conductivity

The conductivity variation for different frequency ranges of pure PUD/PCL blend and with different weight percentage of Fe_2O_3 nanoparticles loaded nanocomposites over the temperature of 30-70°C are displayed in figure 3. Pure polyurethane electrical conductivity is between 10^{-12} to 10^{-8} S/m [21]. The reported conductivity value for polycaprolactone is 1.86×10^{-11} Scm^{-1} at room temperature [22]. But the electrical conductivity of iron oxide at room temperature is found to be in the order of 10^{-2} S/cm [23]. It can be seen; the conductivity of blend was found to be 4.215×10^{-6} to 6.38×10^{-6} S/cm. For Fe_2O_3 nanoparticles loaded composites, the conductivity values increased from 3.13×10^{-5} to 4.69×10^{-5} S/cm for different frequency ranges selected with temperature. Since the conducting polymer nanocomposites electrical conductivity variations depend on the conductivity of the conducting polymer matrix and the filler particles [24.25]. Since the temperature increases, the charge carriers are thermally introduced and the free volume increase, which in turn results more empty places for the motion of ions, and hence increases the conductivity [17].

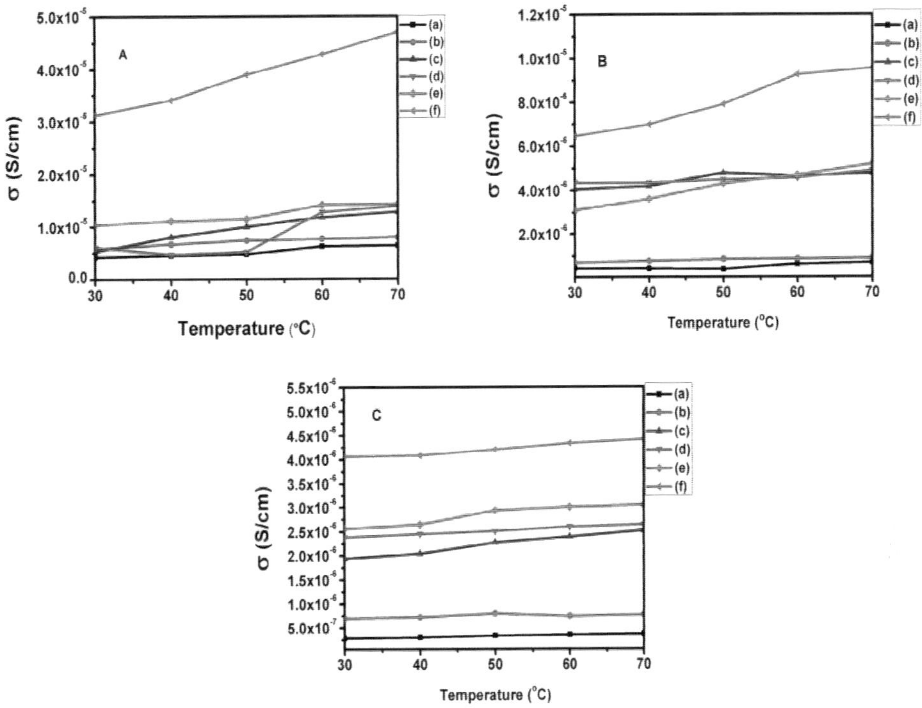

Figure 3. (A-C) Conductivity of frequencies 1 KHz, 10 KHz, 100 KHz for nano iron oxide composites. (a) Blend (b) 0.5 wt% Nano iron oxide composite (c) 1.0 wt% nano iron oxide composite (d) 1.5 wt% nano iron oxide composite (e) 2.0 wt% nano iron oxide composite (f) 2.5 wt% nano iron oxide composite.

3.4 Cyclic voltammetry analysis

Cyclic voltammetry for Polyurethane diol/Polycaprolactone polymer blend and Fe_2O_3 nanoparticle filled nanocomposites were studied in the potential applied between -0.8V to 1.2V verified at 50mVS^{-1} scan rates with 1 M KOH, and are presented in figure 4. The anodic peaks and cathodic peaks were observed for PUD/PCL-Fe_2O_3 nanocomposites. The cathodic potential value for pure blend is 0.117V. But for nanocomposites of higher Fe_2O_3 nanoparticles concentrations viz., 0.5, 1.0, 1.5, 2.0 and 2.5 wt %, reduction potential values are less positive at -0.378V, -0.379, -0.381V, -0.385, and -0.390 respectively. The anodic oxidation peak value for 1.0, 1.5, 2.0 and 2.5 wt % loaded composites, the values are -0.666V, -0.669V, -0.678V, and -0.679V. From the peak values, the current of the nanocomposites increased from the lower concentration to higher concentration by the loading of iron oxide nanoparticles added. This improved electrochemical behaviour can be attributed to the electrical conductivity of the Fe_2O_3 nanocomposites on the electrode surface [26]. The electrochemical response revealed the ability of the material to perform as a potential alternative promoter and base for catalyst in fuel cells.

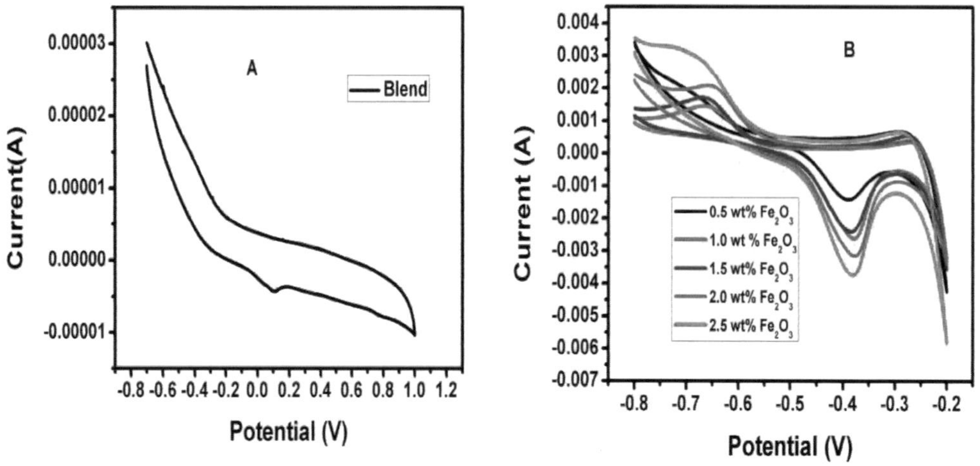

Figure 4. Cyclic Voltammetry curves of various nano iron oxide composites (A) Pure blend (B) 0.5-2.5 wt% nano iron oxide filled composites.

4. Conclusion

The temperature dependence analysis showed that at low temperature, the dielectric constant was low and at higher temperature dielectric constant was increased. As the temperature was increased, the size of grains becomes more that results in the rise of dielectric constant value. Frequency dependence dielectric constant graphs showed that when the frequency was increased, the dielectric constant value decreased. The space charges orient themselves along the applied field, while the electric field was applied. When the applied field frequency increases, polarization and also dielectric constant values decreases. The conductivity value for nanocomposites increased than the pure blend for different frequency ranges selected for different temperature. This can be attributed to the fact that the conducting polymer nanocomposites electrical conductivity changes depend on the conductivity of the filler particles into conducting polymer matrix. Cyclic voltammetry studies shown that the electrochemical response of the material to perform as a potential alternate catalyst. This novel polymer nanocomposites prepared can be utilized for various field of electronics and fuel cells.

Acknowledgement

We acknowledge IIT Bombay, Cochin University, S.T.Hindu College, Nagercoil and Noorul Islam University, Kumarakovil for their technical support.

References

[1] Guelcher SA 2008 Biodegradable Polyurethanes: synthesis and applications in regenerative medicine *Tissue eng part b rev* **14** 3–17.
[2] Nair LS and Laurencin CT 2007 Biodegradable Polymers as Biomaterials *Prog Polym Sci* **32** 762–798.
[3] Santerre JP, Woodhouse K, Laroche G and Labow RS 2005 Understanding the biodegradation of polyurethanes: from classical implants to tissue engineering materials *Biomaterials* **26** 7457–7470.
[4] Mukherjee K, Tribedi P, Chowdhury A, Ray T, Joardar A, Giri S and Sil AK 2011 Isolation of a Pseudomonas Aeruginosa Strain from Soil that can degrade Polyurethane diol *Biodegradation* **22** 377-388.

[5] Rhee S-H 2004 Bone – like apatite- forming ability and mechanical properties of poly(epsilon-caprolactone)/silica hybrid as a function of poly(epsilon-caprolactone) content *Biomaterials* **25** 1167 – 1175.
[6] Misra AN, Hussain MRA, Fiddian NJ 2003 Newton G.The role of the posterior cruciate ligament in total knee replacement *J. Bone Joint Surg (Br)* **85** 389-392.
[7] Ajili SH, Ebrahimi NG, Ansari M 2008 Rheological study of segmented polyurethane and polycaprolactone blends *Rheo. Acta* **47** 81-87.
[8] Bødker F, Hansen MF, Bender Koch C and Morup S 2000 Particle interaction effects in antiferromagnetic NiO nanoparticles *J Magn Mag Mater* **221** 32-36.
[9] Tartaj P, Morales MD, Veintemillas-Verdaguer S, Gonzalez-Carreno T and Serna CJ 2003 The preparation of magnetic nanoparticles for applications in biomedicine *J Phys D* **36** 182-197.
[10] Sun C, Lee JSH, Zhang M 2008 Magnetic nanoparticles in MR imaging and Drug Delivery *Adv. Drug Deliv Rev* **60** 1252–1265.
[11] Fan L, Dang Z, Nan CW, Li M 2002 Thermal, electrical and mechanical properties of plasticized polymer electrolytes based on PEO/P (VDF-HFP) blends *Electrochim. Acta* **48** 205-209.
[12] Ramesh S, Winie T and Arof AK 2007 Investigation of mechanical properties of polyvinyl chloride-polyethylene oxide (PVC-PEO) based polymer electrolyte for lithium polymer cells *Eur Polym J* **43** 1963-1968.
[13] Han H-S, Kang H-R.,. Kim S-W,.Kim H-T 2002 Phase Separated Polymer Electrolyte Based On Poly(Vinyl Chloride)/Poly(Ethyl Methacrylate) Blend *J. Power Sources* **112** 461-468.
[14] Huo L, Li W, Lu L, Cui H, Xi S, Wang J, Zhao B, Shen Y and Lu Z 2000 *Chem. Mater* **12** 790-794.
[15] Suhasini A, Vinod Kumar KP and Maiyalagan T 2016 Synthesis, thermal and magnetic behavior of iron oxide – polymer nanocomposites *Sci Eng Compos Mater* **25** 189-195.
[16] Rahman, SA 2006 Temperature, Frequency and Composition Dependence of Dielectric Properties of Nb Substituted Li-ferrites *Egypt J Solids* **29** 131-140.
[17] Mohapatra SR, Thakur AK, Choudhary RNP 2008 Studies on dielectric properties of a conducting polymer nanocomposite system *Indian J.Eng.Meter.Sci* **15** 347-351.
[18] Abdullah OG, Jamal GM, Tahir DA, Saeed SR 2011 Electrical Characterization of Polyester Reinforced by Carbon Black Particles *Int. J. Appl. Phys Math* **1** 101-105.
[19] Akram M, Javed A, Rizvi, TZ & Turk 2005 Dielectric Properties of Industrial Polymer Composite Materials *J Phys* **29** 355-362.
[20] Fonseca CP, Rosa DS, Gaboardi F and Neves S 2006 Development of a biodegradable polymer electrolyte for rechargeable batteries *J Power Sources* **155** 381-384.
[21] Athanasopoulos N, Baltopoulos A, Matzakou M, Vavouliotis A and Kostopoulos V 2012 Electrical conductivity of polyurethane/MWCNT nanocomposite foams *Polym Compos* **33** 1302-1312.
[22] Woo HJ, Majid SR, Arof AK 2011 Conduction and thermal properties of a proton conducting polymer electrolyte based on poly(ε-caprolactone) *Solid state Ionics* **199-200** 14-20.
[23] Varshney S, Ohlan A, Jain VK, Dutta VP, Dhawan SK 2014 In situ synthesis of polypyrrole –γ –Fe_2O_3–fly ash nanocomposites for protection against EMI pollution *Ind Eng Chem Res* **53** 14282-14290.
[24] Wang Y-G, Xia Y-Y 2006 Electrochemical capacitance characterization of NiO with ordered mesoporous structure synthesized by template SBA-15 *Electrochim. Acta* **51** 3223-3227.
[25] Suhasini A, Vinod Kumar KP 2019 Investigation of polyurethane diol/polycaprolactone: Nickel oxide nanocomposites *J Thermoplast Compos Mater* DOI: 10.1177/0892705719835288.
[26] Lorkit P, Panapoy M and Sapabutr BK 2014 Iron Oxide –based supercapacitor from Ferratrane precursor via sol-gel-hydrothermal process *Energy Procedia* **56**, 466-473.
[27] Suhasini A, Vinod Kumar KP 2019 Solubility and swelling studies of polymer hybrid- nickel oxide nanocomposite,*J Indian Chem Soc* **96** 7-8.

IOP Conf. Series: Materials Science and Engineering **983** (2020) 012010 doi:10.1088/1757-899X/983/1/012010

Experimental investigation on partial replacement of fine aggregate by the effect of Coal bottom ash in concrete

Devananth R[1], Karthikeyan V[1], Sujatha S[1], Anto J [1],

[1] K. Ramakrishanan College of Technology, Trichy, Tamilnadu, India.

devananth.r@gmail.com, karthiathurai@gmail.com sujalalit@gmail.com , antojebas@gmail.com,

Abstract. The fine aggregate is one of the most important constituent in concrete. It will enhance the property like strength, thermal and elastic characteristic of concrete. Owing to minimize the use of "fine aggregate in concrete, the considered amount" of its weight has to be replaced with some other alternative material. Performance of the concrete becomes economically feasible when coal ash powder is used for concrete. "In this investigation, fine aggregate is replaced" with coal bottom ash in different proportions like 20%, 30% and 50%. The strength properties like compression and split tension were compared with two different days of curing. The result reveals the increased in compressive strength was found to vary from 13.2% to 45.57% and maximum split tensile strength was from 12.43% to 54.91% in 28 days of curing the concrete.

1.Introduction

The strength characteristics of concrete will always depend upon the different materials used for construction. Aggregate used in concrete to enhance the strength, thermal and elastic property. In this project, coal bottom ash powder is taken as alternate material for fine aggregate. The coal ash is replaced at different levels 20 % 30 % and 50 % by weight of fine aggregate in concrete. The results were analyzed with the control specimen with no replacement. According to P.Chindaprasirt et al[2] (2004) the effect of coal bottom ash will enhanced the various properties of concrete such as workability, bleeding, setting times, compressive strength, split tensile strength, flexural strength, shrinkage etc.

2. Material Used and Description of Materials

All the materials used in this project are accordingly to I.S. specifications

2.1 *Cement*

The cement which we had used in this work is43 grade of Ordinary Portland Cement conform according to IS: 8112-1989. And the specific gravity was found using pycnometer bottle is 3.15

2.2 *Water*

Drinking water available in Concrete Laboratory was used for casting of all the specimens. Based on the requirements of IS: 456-2000 the quality of water is tested and satisfies.

2.3 *Fine Aggregate*

The fine aggregate is brought from Cauvery River sand, Trichy. It has a specific gravity of2.72. Sieve analysis for fine aggregate was conducted and the sand used conforming Zone III grading. The fine aggregate had sieved which passes through 4.75 mm I.S sieve.

2.4 *Coarse Aggregate*

The coarse aggregate used was hard broken granite stone. The coarse materials was sieved which is passes through 12.5 mm and retain on 4.75 mm IS sieves were used for this investigation.

2.5 *Coal Bottom Ash*
The coal Bottom ash as an alternate materialis from Thermal power station at Allundur, near Trichy. According to M. Singh et al[1] (2013) the coal bottom ash is mainly comprise of fused coarse ash particles which is of fined mix powder is under Class C type of fly ash. The specific gravity was found to be 2.7. The physical property like Grey color and angular texture in shape and of well graded sand size materials and found to be 28 % of dry weight ware observed.

3. Design Mix Proportion
In this work, M25 grade of concrete is taken as a mix proportion. Mix proportion which we adopt has water cement ratio of 0.36 under M25 grade of concrete is 1:1:2. In this fine aggregate weight is replaced by 20%, 30%, and 50 % of coal bottom ash.

3.1*Tested Specimens*
In this project the strength properties like Cube Compression test and split tensile test is undergone. So totally two numbers of dimension (300mm height and 150 mm diameter) cylinder and dimension (150mm x 150mm x 150mm) cubes were tested. And it is examine separately for minimum 7 days and maximum for 28 days of curing.

Table 1: Details of the casted specimens

Mix No	Percentage Replacement of Coal Bottom Ash	Number of Casted cubes		Number of Casted cylinder	
		7 days	28 days	7 days	28 days
A1	0%	2	2	2	2
A2	20%	2	2	2	2
A3	30%	2	2	2	2
A4	50%	2	2	2	2

4. Experimental Investigation
For any successful project work, numerous testing methods have to be performed and analyze before coming into the final discussion. The various tests to be performed for the present work are.

1. Compression Test on Cubes.
2. Split Tensile Test on Cylinders

4.1 Test Programme
The above two mentioned tests are conducted on M25 grade concrete with different proportions of Coal Bottom ash powder.

4.2 Cube Compression Test
Specimen used for the test is of standard dimensioncubes were casted, cured and undergone through Compression Testing Machine of capacity 450 kN. The compression test was carried out with uniform stress of 14 N/sq mm / minute after the specimen had been centered in the testing machine.

IOP Conf. Series: Materials Science and Engineering **983** (2020) 012010 doi:10.1088/1757-899X/983/1/012010

4.3Split Tensile Test

Cylindrical specimen of standard size of (diameter 150 mm and height 300 mm), is kept horizontally under Compression Testing Machine with the maximum load is applied throughout the specimen until the failure is observed through the surface.

5. Result and Discussion

This experimental test of different proportion of coal bottom ash is made to replace for fine aggregate shows that the increasing in strength like compression and split tension occur in maximum percentage of replacement.

5.1 Cube Compression Test

5.1.1 Effect on 0% Replacement

The Cube Compressive strength for 0% replacement coal bottom ash for fine aggregate shows that 13.4 $N/mm^2$and 19.09 $N/mm^2$for different days of curing.

Table 2: Cube compressive strength of concrete

Mix No	Cube Compressive strength N/mm^2 at 7 days	Cube Compressive strength N/mm^2 at 28 days
A1	13.4	19.09
A2	13.6	23.73
A3	13.8	24.2
A4	14.3	27.79

5.1.2 Effect on 20% Replacement with Coal Bottom Ash

When 20% replacement of coal ash made in concrete reveals that increasing its compressive strength by 1.49% for 7 days and 13.82 %for 28 days.

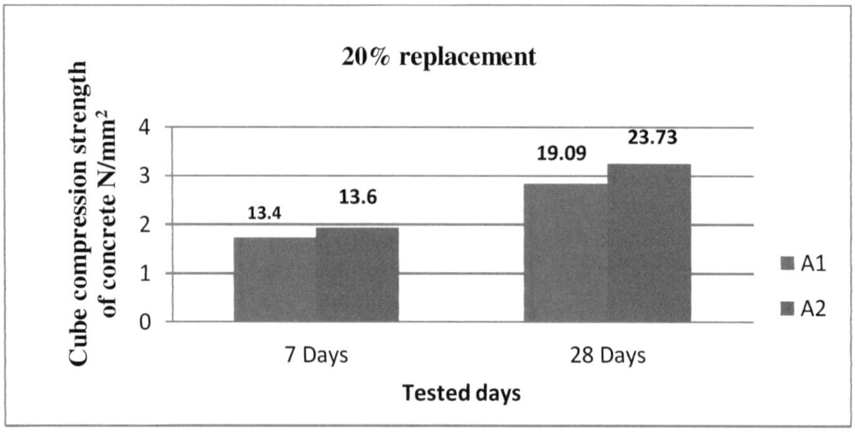

Figure 1. Cube Compressive strength for 20% replacement of Coal bottom ash

5.1.3 Effect on 30% Replacement

When 30% of replacement of coal bottom ash is made for fine aggregate shows that increasing its strength by 2.98% for 7 days and 27.02 %for 28 days.

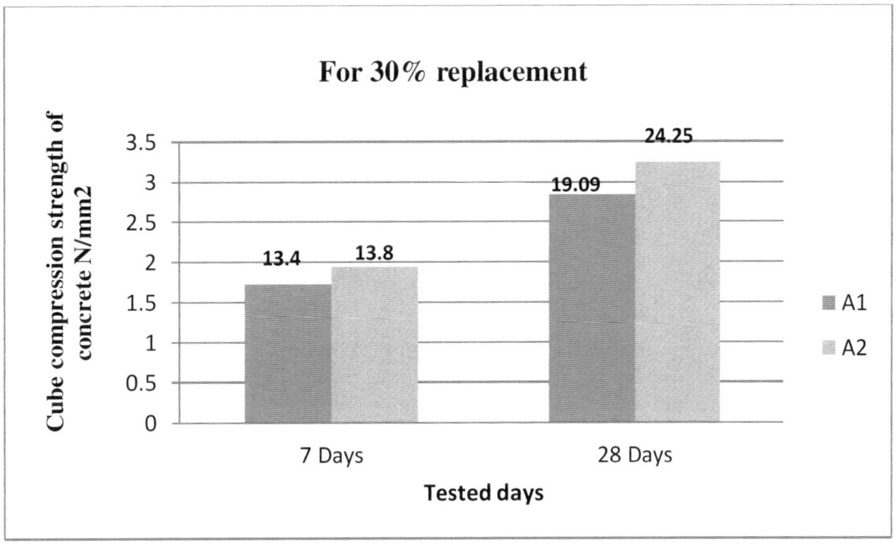

Figure 2. Cube Compressive strength for 30% replacement of Coal bottom ash

5.1.4 Effect on 50% Replacement with Coal Bottom Ash

Compressive strength for 50% replacement of fine aggregate was found to be increasing its strength by 6.71% for 7 days and 45.57 %for 28 days.

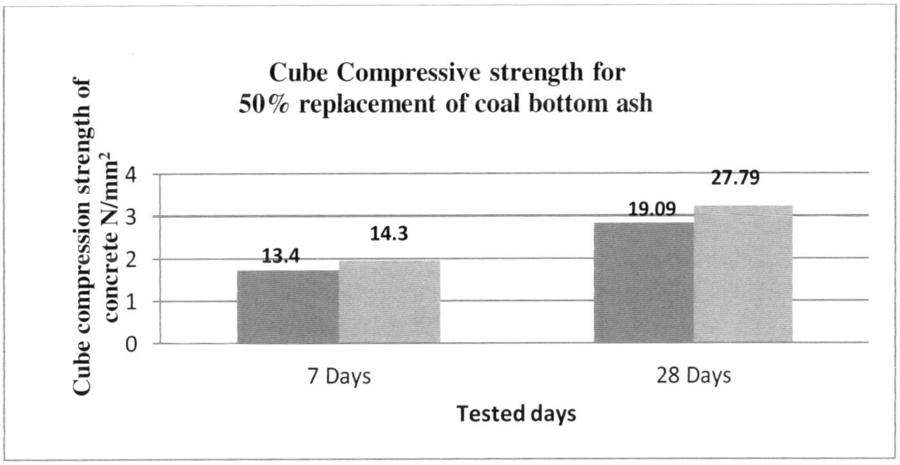

Figure 3. Cube Compressive strength for 50% replacement of Coal bottom ash

Table 3. Percentage increase in compressive strength

Mix No	% Increase in strength at 7 days (N/mm²)	% Increase in strength at 28 days (N/mm²)
A1	0	0
A2	1.49	13.82
A3	2.98	27.02
A4	6.71	45.57

5.2 *Split Tensile Test*

5.2.1 *Effect on 0% Replacement*

Split tensile strength for 0% replacement shows that 1.73 N/mm^2 and 2.83 N/mm^2 for7 and 28 days respectively

Table 4. Details of split tensile strength

Mix No	Split tensile strength (N/mm²)	Split tensile strength (N/mm²)
A1	1.73	2.83
A2	1.94	3.24
A3	2.16	3.57
A4	2.68	3.93

5.2.2 *Effect on 20% Replacement*

For 20% replacement of coal bottom ash for fine aggregate shows that increase its strength by 12.23 % for 7 days and 14.48% for 28 days respectively.

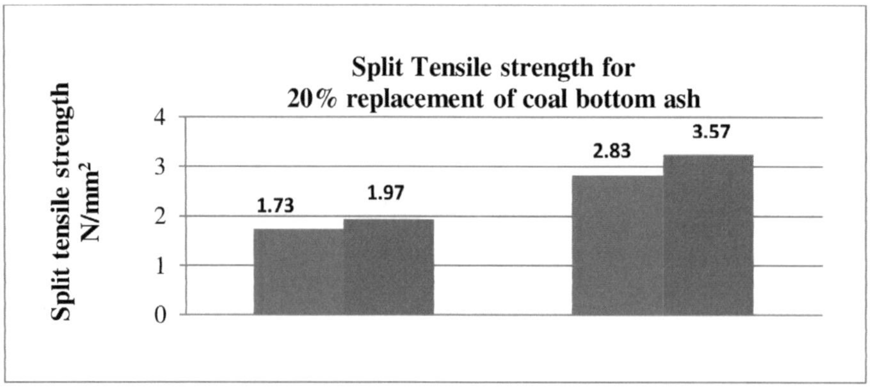

Figure 4. 20% replacement Split tensile strength

IOP Conf. Series: Materials Science and Engineering **983** (2020) 012010 doi:10.1088/1757-899X/983/1/012010

5.2.3 *Effect on 30% Replacement*
The split tensile strength for 30% replacement shows that increase its strength 24.85 % for 7 days and 26.14 % for 28 days.

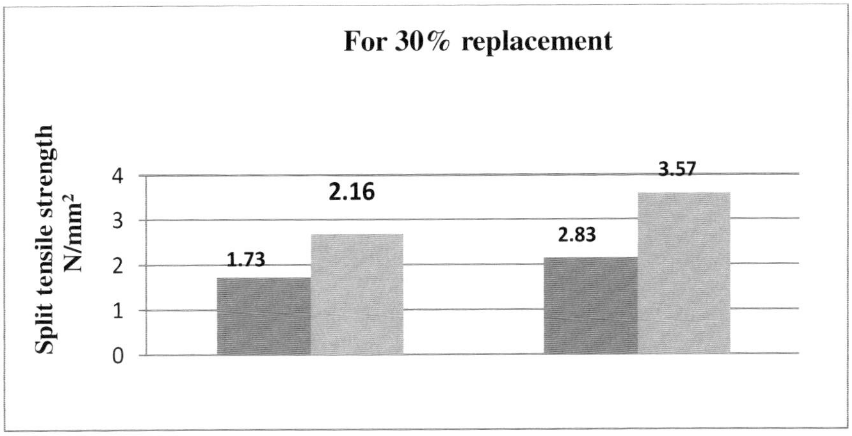

Figure 5. Split tensile strength for 30% replacement

5.2.4*Effect on 50% Replacement*
When 50% of coal bottom ash is made replacement in concrete shows that increase its tensile strength by 54.91 % for 7 days and 38.86 % for 28 days.

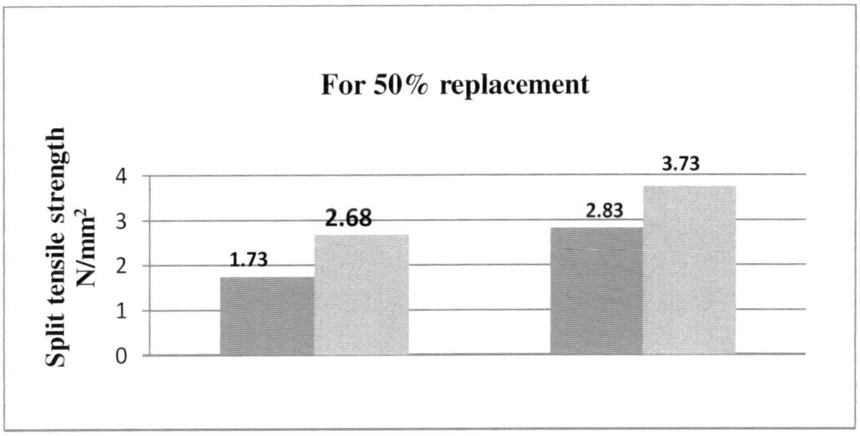

Figure 6. Split tensile strength for 50% replacement of coal bottom ash

5.2.5 *Percentage increase in Split tensile Strength N/mm²*

Table 5. Total percentage increase in Split tensile strength

Mix No	% Increase in Split Tensile test 7 days	% Increase in Split Tensile test 28 days
A1	0	0
A2	12.43	14.98
A3	24.85	26.14
A4	54.91	38.86

6. Conclusion

The test results reveals that on partial replacement of Coal bottom ash powder with fine aggregate shows the maximum compressive strength of 14.3 N/mm^2 for 7 days and 27.79 N/mm^2 for 28 days and it leads to improvement in percentage increase of its strength by 6.71 % and 45.57 % for 7 days and 28 days respectively and for split tensile strength it was observed that 2.68 N/mm^2 for 7 days and 3.93 N/mm^2 for 28 days, which develop the improvement in percentage increased in strength of 54.91 % for 7 days and 38.86 % for 28 days. Therefore it concludes that maximum replacement of coal bottom ash powder in concrete will improves the strength characteristics of concrete.

7.References

[1]. Singh Mand Siddique R 2013 Effect of coal bottom ash as partial replacement of sand on properties of concrete Resources *Conservation and Recycling* **72**20–32.

[2]. ChindaprasirtP, HomwuttiwongS, and SirivivatnanonV 2004 Influence of coal bottom ash fineness on strength, drying shrinkage and sulfate resistance of blended cement mortar *Cement and Concrete Research*, **34**, 1087–1092,2004.

[3]. ParkJ Y 2014 Assessing determinants of industrial waste reuse: The case of coal bottom ash in the United States *Resources, Conservation and Recycling***92** 116–127.

[4]. Harish Kumar and Anandh S 2015Bchavior of Combination of Coconut Fiber and Recron Fiber in Concrete *International Journal of Research in Engineering and Technology***04**206-211.

[5]. IS 10262: 2009 'Indian Standard Code of Recommended guidelines for concrete Mix Design', *Bureau of Indian Standards*, New Delhi.

[6]. IS 383: 1970 'Indian Standard Code of Coarse and Fine aggregates from Natural Sources for concrete', *Bureau of Indian Standards*, New Delhi.

[7]. IS 456: 2000 'Indian Standard Code of Plain and Reinforced Concrete code of Practice', *Bureau of Indian Standards*, New Delhi.

[8]. Understanding and Using Reflective Insulation, Complied by Reflective Insulation Manufactures Association, 2nd edition

[9]. Chudley R. and Greeno R., "Construction Technology", 4th edition, U. K. 2005, PP., 427-42-34

[10]. Sulphate resistance of blended cements containing flyash and rice husk ash – P.Chindaprasirt, P.Kanchanda – Construction and Building Materials 21 (2007) 1356-1361.

Experimental Study on the Presence of Mineral Admixtures and Steel Fiber on the Elastic Properties of Self-Compacting Concrete (SCC)

S.Vijaya Kumar[1*], B L P Swami[2], B.Dean Kumar
[1]Department of Civil Engineering,Vasavi College of Engineering,Hyderabad (Telangana), INDIA
[2]Professor, Department of Civil Engineering ,MCET, Hyderabad (Telangana), INDIA
[3]Professor, Department of Civil Engineering ,JNTUH, Hyderabad (Telangana), INDIA,
*corresponding mail:**s.vijayakumar@staff.vce.ac.in**

ABSTRACT: The elastic properties like young's modulus and poison's ratio are critical issues in the design of concrete structures. This technical paper focuses on the influence of the mineral admixtures like flyash and condensed silica fume along with percentage of steel fibers on the compressive strength and elastic properties of self-compacting concrete (SCC). The flow ability, passing ability along with the segregation resistance tests were conducted. Designed the M40 grade of SCC as per American concrete institute (ACI) provisions. The cement is mingle with flyash(FA) at 20% and condensed silica fume(CSF) at 10% as partial replacement by weight. The steel fiber of diameter 1mm and aspect ratio of 40 mixed at 0.2, 0.4, 0.6, and 0.8 percentages by volume of the concrete. The Young's modulus and Poisson's ratio are resolved as per American standard for testing materials (ASTM) specifications. Based on the experimental investigation, conclusions drawn on the contribution of mineral admixtures and steel fiber on the compressive strength and elastic properties of SCC.A mathematical model developed and checked the validity of the equation with the experimental results.

Keywords: Aspect ratio, Condensed Silica Fume, Chemical admixtures, Poisson's ratio, steel fibers, Young's modulus.

1. Introduction

The structural strength of the concrete is mainly dependent upon the mix design, physical, and chemical properties of the ingredients. Compaction and curing also influence the hardened properties. Self-Compacting Concrete (SCC) defined as concrete material which doesn't require any external compaction for formation of dense mix. This type of special concrete is mainly used for placing in dense reinforcement where compaction is difficult. The SCC is also used in casting of thin concrete structural elements like plates and shells. SCC developed in Japan in late 90's. Initially Okamura, Ouchi, and Ozawa investigated this special type of concrete. Later many researchers carried out the experimentation on the rheological behaviour, structural properties etc. The researchers also studied the influence of admixtures on the structural properties along with the addition of fibers in it. Basically concrete is strong in direct compression, by adding fibers in the concrete matrix the tensile and flexural strengths were improved. But the percentage of these fibers in SCC depends upon their physical properties. For steel fibers it is limited to a particular percentage of volume of the concrete, so as to avoid the obstruction of flow ability. By adding admixtures to the concrete, the properties in fresh stage are not only improved, but the strength properties in the hardened stage also enhanced. The chemical admixtures like superplasticizer (SP) improves the flow ability of the concrete. The viscosity modifying agent (VMA) modifies the segregation resistance. The percentage of these admixtures depend upon the properties of the ingredients used in the mix.

2. Literature review

In late 90's Okamura.H, along with his team of scientists developed the ideology of SCC. This type of the concrete developed at that time for better strength and durability by using supplementary

IOP Conf. Series: Materials Science and Engineering **983** (2020) 012011 doi:10.1088/1757-899X/983/1/012011

cementatious materials (SCM) such as mineral admixtures. As per Okamura mix design is related to fine aggregate to mortar ratio to evaluate compatibility and flow ability. Nan Su et al (1), developed the easy mix design concept, which is based on the ratio of fine aggregate to total aggregate by mass. The authors concluded that the quantity of fine aggregate is more than the coarse aggregate in the complex of concrete, which can help the passing ability of the concrete through reinforcement bars easily. Okamura and Ouchi (2) reported the current and future perspectives of SCC to the world and given an outline way for the design mix of SCC incorporating the mineral admixtures at optimum dosages to satisfy the needs of the SCC. Turhan Bilir (3) did the experimentation for prognosticate of the elastic modulus for high grade concrete which incorporates the mineral admixtures such as silica fume and ground pumice. The author developed empirical formula for the material constant of the concrete. F. P.Zhou et al (4) investigated the affect of the coarse aggregate on the mechanical properties of the High performance concrete along with the Modulus of elasticity. D.Stefaniuk et al(5) did the experimentation for the determination of elastic properties of self-compacting concrete modified with nano particles. It was developed by multi scale approach.F.Demir(6,7) did the experiments by using the artificial neural network, to prognosticate the modulus of elasticity for normal and high grade concretes. Ilker Bekir et al (8) did an attempt by using the composite material models for the estimation of modulus of elasticity of slag concrete. T.Noguchi et al (9) did the investigation for development of practical equation for the modulus of elasticity of concrete. H.Z.Cui et al(10) gave the analytical models for the compressive strength, and Modulus of elasticity. The authors also investigated the peak strain of structural lightweight aggregate concrete.Vijaya Kumar et al (11) developed the mathematical model for the compressive strength and elastic properties for the triple blended steel fiber self-compacting concrete. K.M.Lee,J.H.Park(12)did the investigation on the development of the numerical model for modulus of elasticity of concrete by considering interfacial transition zone.A.H.Gandomi et al(13) extracted the new design equations for modulus of elasticity of concrete using multi expression programming. "European Federation of National Associations Representing for Concrete-specifications for the self-compacting concrete",EFNARC-2002(14) gave guidelines for Self Compacting Concrete. American Concrete Institute ACI-544 1R-960(15) gave the testing procedures for the determination of elastic properties. American Standards Testing of Materials (ASTM C 469-02)(16) gave the testing procedures for the determination of elastic properties,

3. Materials used for Experimentation:

3.1 Cement(OPC): 53 grade confirming to the IS:269-2019, used and all the properties such as its physical and chemical were examined as per the specifications.
3.2 Fine Aggregate: River sand confirming to Zone –II was adopted. Its modulus of fineness is 2.8.
3.3 Coarse aggregate(CA): Passing through 12 mm sieve and hold on on 10mm sieve used in experimentation, such that this type of the coarse aggregate is very easy to fulfil the rheological requirements. Shape of the coarse aggregate is the combination of rounded and angular. The modulus of the fineness is 6.8.
3.4 Mineral Admixture: F-type of flyash(FA) along with the condensed silica fume(CSF) is blended with the cement, the specific surface areas are 4000 and 150000cm^2/gram respectively. The CSF has non-crystalline structure whereas FA is mostly globular. The CSF and FA has specific gravity 2.2 and 2.3 respectively.
3.5 Chemical Admixture: Glenium B 233 based on poly carboxylic ether (PCE) as superplasticizer along with Glenium-2 as VMA was used for this investigation.
3.6 Steel Fibers: Ordinary steel fiber of 1mm diameter and with an aspect ratio of 40 is used in the present investigations.
3.7 Water: water used for mixing, curing which is free from oil and any soluble salts.

4. Experimentation

4.1 Mix Design and Workability Tests:

M40 grade of SCC is developed based on the simple design approach by Nan Su[1]. The ratio of volume of the fine aggregate to the total aggregate content in the concrete is around 0.50 to 0.57 which depends on the packing factor(PF).In this experimentation, above ratio adopted as 0.62, by various trials to satisfy the needs of the SCC as per EFNARC guidelines. It basically depends upon the properties of the ingredients used in the SCC. Final mix design for the SCC along with the quantity of the materials are given in Tables 1 and 2.

Table 1.M40 Grade SCC-Mix Proportions

S.No	Cement	Fine Aggregate	Coarse Aggregate	Water to cementitious ratio
1	1	2.32	1.38	0.45

Table 2. The Materials Adopted -M40 Grade SCC.

S.No	Material	Quantity for one cu.m
1	Cement	320 Kg
2	Flyash(FA)	92 Kg
3	Condensed Silica Fume(CSF)	46 Kg
4	Fine Aggregate(river sand)	1056 Kg
5	Coarse Aggregate(12mm passing/10mm retained)	628 Kg
6	Water	204 liter
7	Superplasticizer(SP)	234 ml
8	Viscosity Modifying Agent(VMA)	24ml

The workability tests for SCC such as flow ability, passing ability and segregation resistance conducted on slump cone, V-funnel, and L- box respectively. These results are achieved when cement is partially blended with FA, and CSF at 20 and 10 percentages by mass of the cement. The SP and VMA were adopted at 1.0 and 0.15 percent by weight of the cement in the mix consistently for all the percentage of the steel fibers. Steel fibers were added at percentages of 0, 0.2, 0.4, 0.6,and 0.8.Details of the experimental results for the workability of the concrete for critical percentage of steel fiber (0.8) and maximum aspect ratio(40) are given in table 3.

Table 3.Workability Properties of Typical SCC

S.No	Name of the test	units	Experimental value*	Acceptance criteria
1	Slump flow	mille meter	720	650-800
2	V-Funnel	seconds	8	6-12
3	L-Box	ratio	0.90	0.80-1.00

4.2 Number of mixes and specimens cast and tested:

The number of mixes adopted in this investigation were 5. For each mix 3 cubes, 6 cylinders were cast. For the determination of the compressive strength, a total of 15number of cubes (150mm side) and a total number of 30 standard cylinders (150mm diameter and 300mm height) were cast for finding the failure load and the elastic properties. All the cubes are tested for compressive strength after completing 28 days of water curing. The 15 number of the cylindrical specimens are also tested for the compressive strength (failure load on the specimen), remaining 15 number of the cylindrical specimens are tested for the Young's Modulus(E) and Poisson's ratio(v) by applying compressive load at 40% of the failure load of the specimen, as per ASTM standards. The arrangement for measuring the dial gauges in longitudinal direction and lateral directions are shown in Figure 1 for noting the strain in both the directions. The testing setup for the cylindrical specimen is shown in Figure 2.

5. Results and Discussion

5.1 Compressive Strength and Material Constants:

The triple blended self-compacting concrete (TBSCC) with 20% FA and 10% CSF is recorded as 44.50MPa as compressive strength. The percentage of steel fiber is increased from 0.0 to 0.8 percent by volume of the concrete in four combinations(0.2,0.4,0.6,and 0.8) by maintaining maximum aspect ratio (40) in each percentage of steel fiber, it observed that the compressive strength of the triple blended steel fiber self-compacting concrete (TBSFSCC) is increased, the results are given in Table.4.The maximum compressive strength recorded for limiting percentage of steel fiber is 68.50MPa. Its variation shown in Fig 3.

Fig 1.Arrangements of the dial gauges

Fig 2.Setup for the elastic constants.

Table 4.The Compressive Strength for TBSFSCC.(20 % FA and 10 % CSF)

S.No	MIX CODE	Mix Abbreviation	Average Cube Strength in MPa
1	M_1-0.0 SF	Mix 1 for 0.0% of steel fiber	44.50
2	M_2-0.2 SF	Mix 2 for 0.2% of steel fiber	49.80
3	M_3-0.4 SF	Mix 3 for 0.4% of steel fiber	54.00
4	M_4-0.6 SF	Mix 4 for 0.6% of steel fiber	60.75
5	M_5-0.8 SF	Mix 5 for0.8% of steel fiber	68.50

The triple blended Self-Compacting Concrete (with mineral admixtures and no fibers) is noted as 32,670MPa as its Young's Modulus value, which is 3.5% more than the standard Young's modulus value for M40 grade of normal concrete(5000 $\sqrt{f_{ck}}$), which is due to the presence of admixtures in the concrete. As observed from the experimental results, with increase in the steel fiber percentages from 0.2 to 0.8, the elastic constant for E-Young's modulus increased where as the Poisson's ratio decreased. There is an increase of nearly 34% for 0.8% of fibers compared to 0% fiber. The reason for increase in the Young's modulus value is due to the presence of steel fiber ,which makes denser compare to SCC without fiber. The strains are less in SFSCC due to locking of the fiber with ingredients in it enhancing the strength. But in case of Poisson's ratio, lateral strain is less for SFSCC compared to SCC without fiber. Hence, Poisson's ratio is reduced with the increase in percentage of steel fiber (nearly 30%). The details of the experimental results shown in Table 5, and its variations shown in Fig 4 and 5.

Table 5. The Elastic Constants for the Steel fiber Self-Compacting Concrete (SFSCC) with triple blending. (20 % FA and 10 % CSF)-Aspect Ratio of steel fiber is 40.

S.No	Mix	E-Young's modulus in MPa	ν - Poisson's ratio
1	M_1-0.0 SF	32,670	0.152
2	M_2-0.2 SF	35,500	0.133
3	M_3-0.4 SF	37,200	0.130
4	M_4-0.6 SF	42,000	0.117
5	M_5-0.8 SF	44,900	0.107

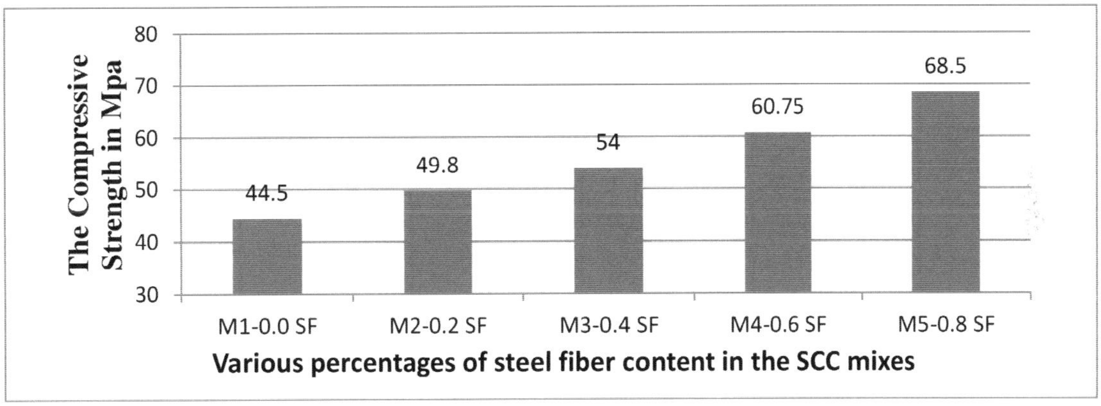

Fig 3. The compressive strength variation for TBSFSCC

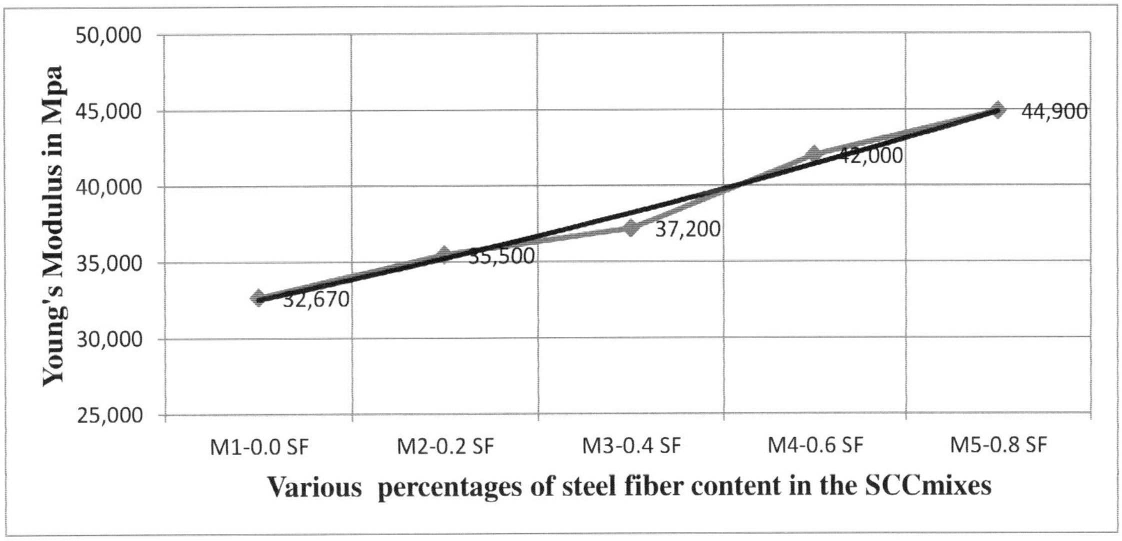

Fig 4. The Young's Modulus (E) variation in TBSFSCC with constant aspect ratio.

5.2 Correlation with mathematical models.

1)The mathematical model for the Young's Modulus for the above SFTBSCC with aspect ratio of 40 is given by

$$E=30014\ e^{0.0804\ (p)}$$

E- Young's Modulus in MPa;

p- percentage of steel content.

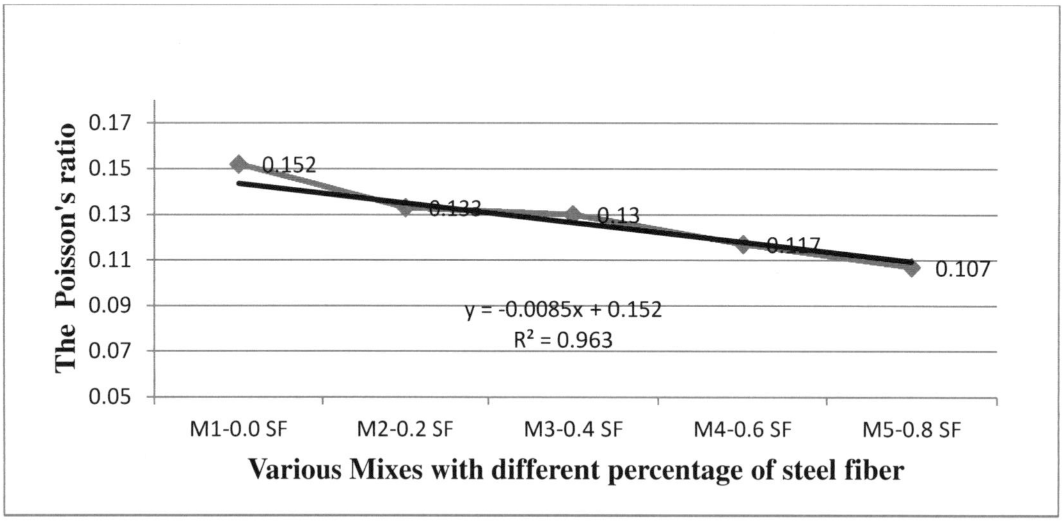

Fig 5. The Poisson's ratio variation in TBSFSCC with constant aspect ratio.

2) The mathematical model for the Poisson's Ratio (**v**) for the above TBSFSCC with aspect ratio of 40 is given by

$$v = -0.0085\ (p) + 0.152$$

v -Poisson's ratio;

p-percentage of steel content.

6. Conclusions

The following conclusions were arrived at based on the experimentation done,

1. Optimum dosages are of the admixtures for getting better rheological characteristics of the Self-Compacting Concrete are 20% flyash, and 10% Condensed silica fume, due to calcium-silicate-hydrate(C-S-H) formation in the cement mortar.
2. The maximum percentage of steel fiber content in the SCC mix is limited to 0.8% of volume of the concrete. Beyond this percentage, the flow properties are interfered. The aspect ratio of the steel fiber is limited to 40 to facilitate smooth flow of S.C.C.
3. The Compressive strength of the basic SCC mix (without fiber) is increased by nearly 50% with 0.8% fiber content in the mix.
4. The Young's Modulus of the TBSCC is 3.5 % more than the Normal concrete of same grade, due to the affect of the mineral admixtures.
5. The steel fiber percentage increased from 0.0% to 0.8 %, the Young's Modulus [E] is enhanced by 34% compared to same SCC without steel fiber content due to more stiffness of the concrete.

6. The Poisson's ratio has decreased by 30 % for the optimum fiber content due to the increase in steel fiber content of SCC due to its more stiff behaviour.
7. The mathematical models for this concrete has showed good correlation to the elastic properties.

Acknowledgment

The authors expressed their gratitude to the administration of the Vasavi College of Engineering, for the facilities provided, also thankful to the UGC to the grant released for the experimentation.

References

[1] NanSu, Kung-Chung Hsu, and His-Wan Chai 2001 A simple mix design for Self Compacting Concrete *Elsevier Journal for Cement and Concrete Research 31 12 1799-1807.*
[2] Okamura,H and Ouchi, 1999 Self-Compacting Concrete-Development present and future *First international symposium on Self Compacting Concrete 3-14.*
[3] TurhanBilir 2016 Investigation of performances of some empirical and composite models for predicting the modulus of elasticity of high strength concretes incorporating ground pumice and silica fume *Jounal of Construction and Building Material* 127 850-860.
[4] Zhou FP, Lydon.F D, Barr BIG 1996 Effect of coarse aggregate on Modulus of elasticity and compressive strength of high performance concrete, Cement *and Concrete Res* 25 177-186.
[5] Stefaniuk D, Niewiadomski P, Musial M, Lydzba D 2019 Elastic Properties of self-compacting concrete modified with nanoparticles:MultiscaleApproach *Archives of civil and mechanical engineers 19 1160 to 1162.*
[6] FuatDemir 2008 Prediction of elastic modulus of normal and high strength concrete by artificial neural networks *Journal of Construction and Building Material* 22 1428 to 1435.
[7] FuatDemir 2005 A new way of prediction Modulus of elasticity of normal and high strength concrete *Cement and Concrete Res* 35 1531 to 1538.
[8] IlkerBekir T, Turhan,B, AhmetRaif B 2010 Estimation of modulus of elasticity of slag concrete by using composite material models *Construction and Building Material* 24 741 to 748.
[9] Noguchi T, Tomosawa F, Nemati KM, Chiaia BM, Fantilli AP 2009 Practical equation for Modulus of elasticity of concrete *ACI ,Structural Journal* 106 690 to 696.
[10] Cui HZ, Lo TY, Memon SA, Xing F, shi X 2012 Analytical Model for the compressive strength, Modulus of elasticity and peak strain of structural lightweight aggregate concrete *Construction and Building Material* 36 1036 to 1043.
[11] Lee KM, Park JH 2008 A numerical model for modulus of elasticity of concrete considering interfacial transition zone *Cement and Concrete Res* 38 396 to 402.
[12] VijayaKumar S, DeanKumar B, Swami BLP 2020 Mathematical Model for the Compressive strength and Elastic properties of the triple-blended steel fiber self-compacting concrete from the experimental investigation *Advances in Structural Engineering, ,Springer publications* 217 to 229.
[13] Gandomi AH, Faramarifar P, GhanadRezaee, Asghari A, Talatahari S 2015 New design equations for Modulus of elasticity of concrete using multi expression programming *Journal of Civil Engineering Management,* 21 761 to 774.
[14] EFNARC-2005,"*Specification and Guidelines for Self Compacting Concrete*"
[15] ACI-544 1R-960,"*State of the Art Report on Fiber Reinforced Concrete Using Recron 39 fibers"*
[16] ASTM C-469-02,"*Standard test method for Static Modulus of Elasticity and Poisson's ratio of concrete under compression".*
[17] BIS-10262-2019- *Indian Standard Concrete Mix Design – Guidelines- Bureau of Indian Standards,NEW DELHI 110002.*
[18] BIS-383-2016- *Indian Standard Coarse & fine aggregate for concrete - Specifications- Bureau of Indian Standards.*
[19] BIS:516 – 1959(Reaffirmed-2004)- *Indian Standard Methods of Tests For Strength of Concrete- Bureau of Indian Standards.*

Land use Land cover studies and its effects on Valuationusing GIS Techniques in Madurai Town Planning Area, Tamilnadu, India

Balaji L[*], **Muthukannan M**[#],

***Balaji L**,Scholar, School of Environmental and Construction Technology, KARE, valuerbalaji@gmail.com

#Muthukannan M, School of Environmental and Construction Technology, KARE, m.muthukannan@klu.ac.in

Abstract Landuse land cover mapping practices over period in Madurai district varies due to increase of urban and rural migration. Valuation of land in urban areas is very important tool towards economic impact on human communities. Madurai is a temple city located on Vagai river bank which attract most tourists and also increase the urban areas due to various factors such as rural migration, Population increase. With the evidence from Government Land Register (GLR) and Present Market Rate (PMR)Values the Population is directly proportional to the value of land. Population increase last 12 years and built-up areas also increased. This paper studies the urban sprawl using land use land cover changes over the period of 2007 to 2019 between GLR and PMR values as per real estate practices in India. Using GIS techniques the analysis of urban sprawl of the city as well GLR, PMR values relationship by interpolation method to assess the land valuation practices which followed by Government of India.

Keywords: Urbanization; Land use and Land cover;, GIS; Present Market Rate.

1. INTRODUCTION

Landuse is the one of important factor on this world, where how people are utilizing the land by different way such as building, agriculture and other purposes. Land cover is the identified by its cover type of region such as forests and other areas. Land use Land Covers mapped using satellite images by image processing techniques and GIS analysis. Growth of city rapidly changes over the period called urban sprawl which also important factor for planning and real estate land valuation for future progress of communities. To balance the conservation and developmental pressure the analysis of land use is very important. Disturbanceofproductiveland,urbanencroachment,anddepletionofforests are the some of the issues that drive land use studies (MalleswaraRao et al. 2012). Valuation is estimation of land or property. Land use data will helpful towards theassess the valuation of land guidance with GLR (Government Land Register) and PMR (Privilege Market Rate) (Maulik and Chakraborty. 2011). The expansion of urban area are predicted and the reason for urbanization with land valuation practices are analyzed for future real estate development and land assessment towards the guideline practices. GIS may be one of useful technique to identification change detection of land use which indirectlyhelps to the estimation of land values. In this paper interpolation method was used (Inverse distance weight) technique (Shougeng et. al, 2013) for overlay analysis with land use and urban sprawl. (Mokhamad Suriantoet.al, 2019) in his research explained about the rapid development of Blora District that it affects the distribution of the market value of the land. The results showed that from eight observed variables, six variables that significantly affect the value of land, namely variables of distance to the city centre, distance to the local market, accessibility, land use, availability of facilities and infrastructure, and land location. Study area of the map shown in the fig.1.(Aitkenhead, M. J et.al,2011) stated An evidence integration rule set is applied to evidence lists to produce conclusions of different strength regarding individual classes, and the most likely class identified. The only expert system design implemented currently within the methodology is a neural network model,

IOP Conf. Series: Materials Science and Engineering **983** (2020) 012012 doi:10.1088/1757-899X/983/1/012012

although the system has been designed to accept information from decision trees, fuzzy k-means and Bayesian statistics as well (Sang et al. 2011).(Jannet C. Bencure et.al 2019) holistic and objective way of weighing geospatial factors through expert consultation, legal reviews, and evidence (i.e., news) will provide more realistic results than a regression-based method that does not comprehend valuation factors (i.e., physical, social,economic, environmental, and legal aspects). The analytic hierarchy process (AHP) enables these factors to be included in the model, hence providing a realistic result. The innovative land valuation model (iLVM), developed in this study, is an inclusive approach wherein experts are involved in the selection and weighing of 15 factors through the AHP. The model was validated using root mean squared error (RMSE) and compared with multiple regression analysis (MRA).In ther study, Garmin GPSwas used for capturing the location in various parts of Madurai. The GLR and PMR was collected with the help of GPS for land in Madurai (Longley et al. 2010).The change that has happened over time, landcover maps taken during different years are needed. With this information, people can evaluate the past as well as gain insight into the possible effects of their current decisions before they are implemented (Seto et al. 2011).Bigdeli et.al (2013) new method for classification of hyperspectral data based on a band clustering strategy through a multiple Support Vector Machine system. The proposed method uses the band grouping process based on a modified mutual information strategy to split data into few band groups. After the band grouping step, the proposed algorithm aims at benefiting from the capabilities of SVM as classification method. So, the proposed approach applies SVM on each band group that is produced in a previous step. Finally, Naive Bayes (NB) as a classifier fusion method combines decisions of SVM classifiers. Experimental results on two common hyperspectral data sets show that the proposed method improves the classification accuracy in comparison with the standard SVM on entire bands of data and feature selection methods.

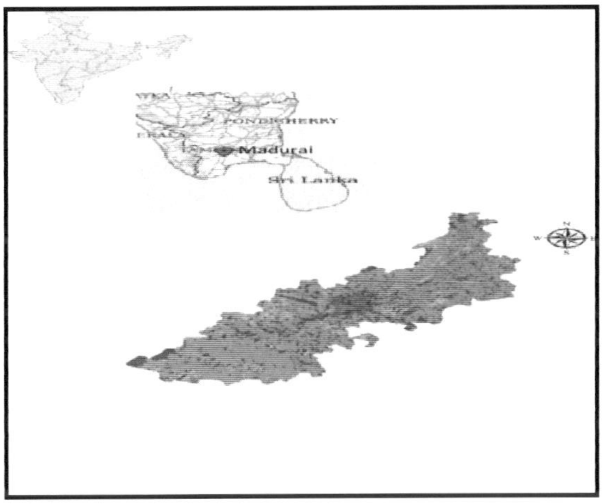

Figure1. Study Area

2. STUDY METHODOLOGY

2.1 Spatial and Field Data Collection:

Landsat Satellite Images are collected between the year 2007 and 2019. Sensors of Landsat7 & 8 ETM, ETM+, TM, and OLI (Source: https://earthexplorer.usgs.gov/) are primary source of this Satellite images. Image resolutions area 30m and 23.5 meter and Metadata information provided as below

Landsat Product Identifier: LE07_L1TP_143053_20070601_20170103_01_T1

IOP Conf. Series: Materials Science and Engineering **983** (2020) 012012 doi:10.1088/1757-899X/983/1/012012

Landsat Scene Identifier: LE71430532007152ASN00
Acquisition Date: 2007/06/01
Scan Line Corrector: OFF
Collection Category: T1
Collection Number: 1
WRS Path 143
WRS Row 053
Date L-1 Generated 2017/01/03
Start Time 2007:152:04:55:31.2928749
Stop Time 2007:152:04:55:58.0469375
Station Identifier ASN
Day/Night Indicator Day

Images are rectified using Georeferencing techniques with ground validated with signature file collected using GPS. Open source software QGIS & Grass used for analyzes the image and derives the land use data. GPS are commonly used for positioning and navigation. GPS data used field data of land or properties collection as one of the important parameter to identify the locations and various attribute information. GPS waypoints are collected with field data sheet to fill up the non-spatial data in various parts of Madurai. Field data sheet contains address and type of built-up such as residential, industry including ward number and streets. Also GLR and PMR values are integrated in the field data sheet for precisely integrate with land use information. .

After Data collecting the data using GIS software points were plotted using lat long information which contains various attribute such GLR, PMR, and others. Old land values are collected from Government agencies for the validation. TheLandsat satellite images was collected from2007 to2019 by USGS Earth Explorer for the research area. The accompanying information were arranged by the computerized Image Processing Software (ENVI) so as to order the pictures utilizing neighborhood calculation and mark set for Level 1 characterization, for example, Settlement (Build-up Area), Forest, Agriculture land, Wasteland including Fallow land and Water bodies by utilizing Unsupervised and Supervised grouping strategies was utilized.Interpolation (IDW) performed using GIS tool to plot the GLR and PMR values in the same study area. Fig.2 shows the methodology of study area.

.

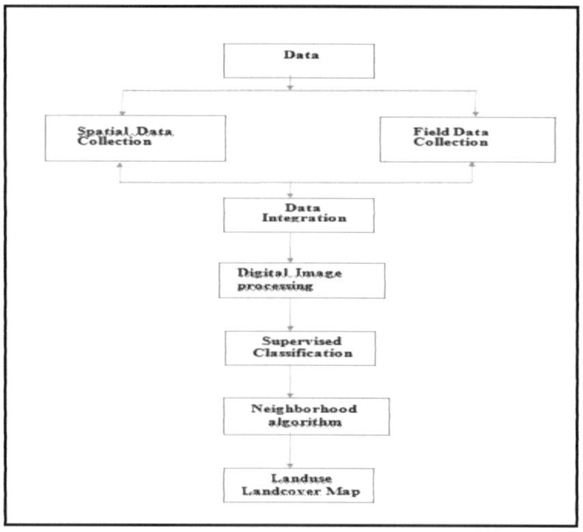

Figure 2.Methodology of GIS work flow

2.2 Data Integration with GIS and Mapping

Various data sources were collated and presented in the form of attributes. After collecting data, was integrated with the GIS software using QGIS. The collected data were tabulated. The tabulated data were provided as input using inverse distance weighted (IDW) technique. With the help of IDW, an imaginary surface was designed with the GLR, PMR values and latitude and longitude details were also viewable and a cluster was formed. The land value can be easily interpreted with the output images. With the help of latitude and longitude (X,Y) and GLR, PMR(Z value), the spatial distribution map was derived from QGIS and also Madurai city area is divided into 3 zones named A, B and C based upon the spatial distribution map of GLR and PMR values. A is defined as area falling under low range lands values when B refers to area falling under medium range lands values and C refers to area falling under high range lands values.

3. SPATIAL ANALYSIS

Thelanduse and landcover for Town Planning Area of Madurai was prepared from satellite image and processed using image processing techniques for the year 2019 shown in Figure 3.

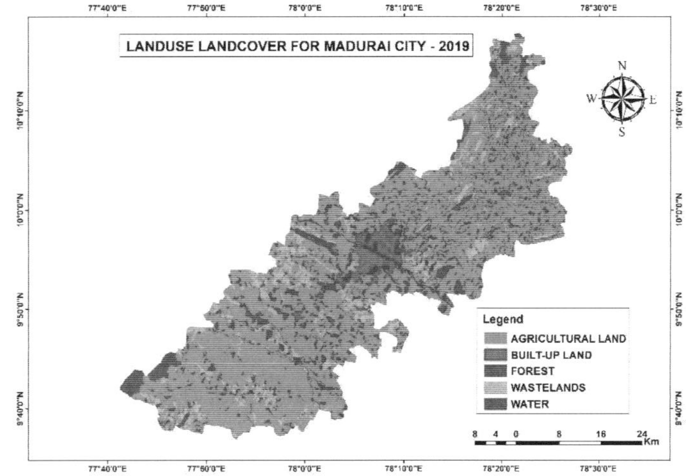

Figure 3. Shows the Land use Land cover map 2019

By using of GIS software,thematic layer land use prepared with field training sets and GIS overlay analysis such as clipping, editing and digitizing. After classification proper land use such as forest, wasteland, settlement area is derived using UTM coordinate system in square kilometer. Maximum area falls under agriculture area shown 877.145 sq.km and minimum area shows forest about 83 sq.km. 2019 image captured during summer season so wasteland projected was high about 800 sq.km.Urban and rural settlement was identified in LPA shows 223 sq.km in the overall area. This settlement area taken for urban sprawl study which was overlaid with GLR and PMR interpolated image for further analysis. Water bodies shows 220.441 sq.km which the source of water for entire study area which also impact of urban sprawl towards LPA. Fig.4 depicts statistics of year 2019 followed by Table 1.

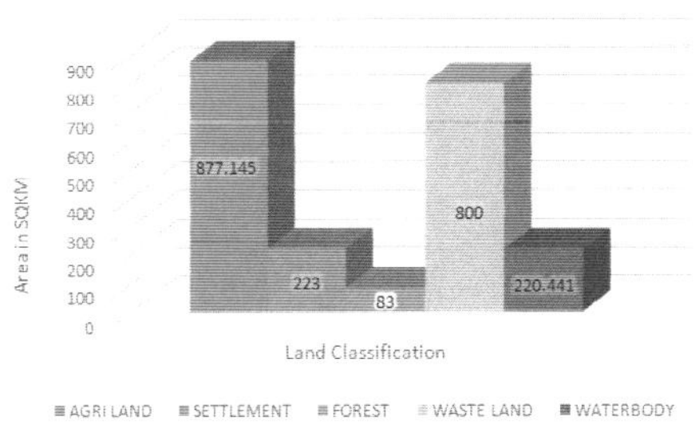

Table 1.Land use Land Cover for year 2019

2019	Area in Sq.km
AGRI LAND	877.145
SETTLEMENT	223

IOP Conf. Series: Materials Science and Engineering **983** (2020) 012012 doi:10.1088/1757-899X/983/1/012012

FOREST	83
WASTE LAND	800
WATERBODY	220.441

Figure 4. 2019 thematic layer statistics

In the Year 2007 land use land cover were mapped using Landsat ETM images acquired, and processed shown in the fig.5. This imagery is cloud free data in the summer season were taken into account for the land use accurate classification with the help of existing field signature training sets. This was used different iterations with K-means method for proper analysis. The settlement area shows less about 148 sq.km compare to 2019. Whereas Wasteland shows maximum area about 1200 sq.km and minimum area is forest which is unchanged. Agriculture and waterbody areas were 536 sq.km and 237 sq.km respectively shown in table 2. Fig.5 shows the map of LULC for 2007 in LPA area of Madurai.

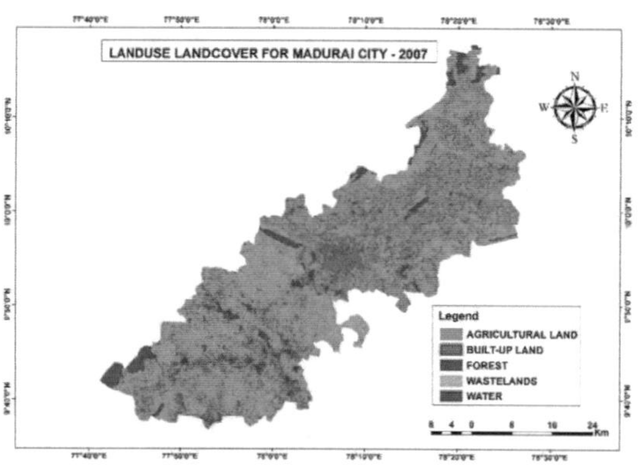

Figure 5. Landuse Land Cover Map for 2007

The agriculture land has been slightly increased from 2007 to 2019 because of varies in acquisition time and cultivation may practice during this period. Agriculture land use increased from 535 sq.km to 877 sq.km. Whereaswasteland decreased is shown in the figure.6

Table 2: Landuse Land Cover for Year 2007

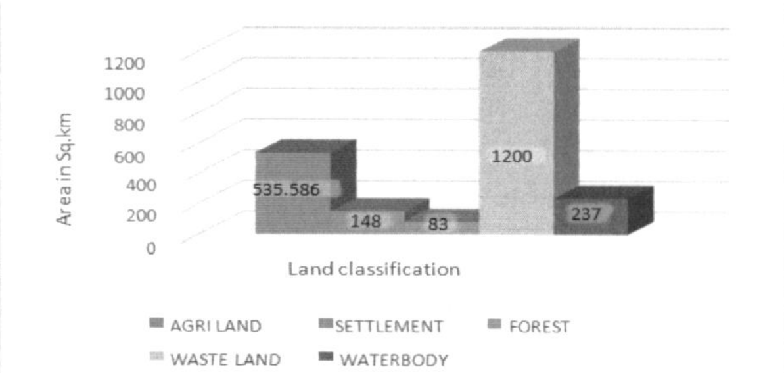

2007	Area in Sq.km
AGRI LAND	535.586
SETTLEMENT	148
FOREST	83
WASTE LAND	1200
WATERBODY	237

Figure 6. 2007 thematic layer statistics

From 2007 to 2019 which shares the landuse with horticulture and created region. Hold RF stays unaltered and Water body (water system source region) somewhat expanded. Table.3 shows the change identification of LULC. From the above Fig. 7, the change discovery of Landuse Land spread between 2007and 2019 shows Agriculture land has expanded 30%. The Settlement territory has additionally filled wide in range from 148 Sq.km to 223 Sq.km. The backwoods stays unaltered. The Barrel land has diminished from 1200 Sq.km to 800 Sq.km. Water bodies has indicated a slight reduction to 220.441 Sq.km from 237 Sq.km in the given table.4

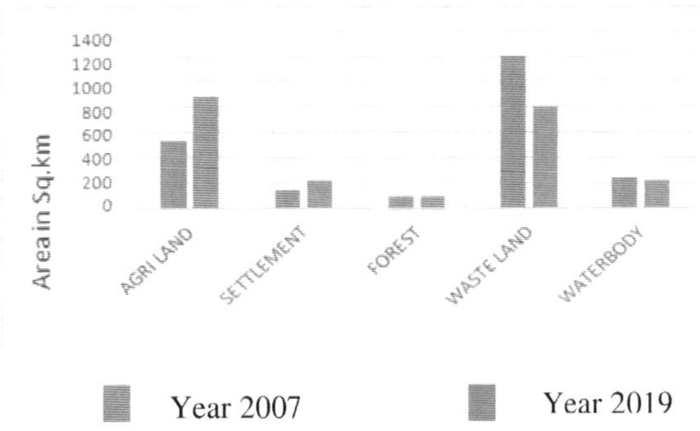

Figure 7. Change detection of LULC 2007 -2019

Table 3. LULC Change detection between 2007-2019

	2007 (in Sq.km)	2019 (in Sq.km)
AGRI LAND	535.586	877.145
SETTLEMENT	148	223
FOREST	83	83
WASTE LAND	1200	800
WATERBODY	237	220.441

Table 4. Comparison between Government Land Register values in 2007 and2019

	A		**B**		**C**	
	Min	**Max**	**Min**	**Max**	**Min**	**Max**
Government Land Register for the year 2007	1.40	317.00	317.00	1240.00	1240.00	2775.00
Government Land Register for the year 2019	19.00	2201.00	2201.00	4000.00	4000.00	5450.00

Table 4 shows GLR factor for A region increased from 1.4 to 19 between 2007 to 2019. The GLR for B and C region increased from 1241 and 2774 to 4000 and 5450 respectively. This averaged value arrives from the IDW methods by spatial distribution overlaid with landuse.

Table 5. Comparison between PMR values in 2007 and2019

	A		**B**		**C**	
	Min	**Max**	**Min**	**Max**	**Min**	**Max**

Present Market Rate for the year2007	10.00	2720.00	2720.00	4350.00	4350.00	5435.00
Present Market Rate for the year2019	120.00	3581.00	3581.00	7040.00	7040.00	8770.00

Table.5, the PMR factor for a district expanded from 10 to 119 inside 12 years. The PMR for B and C locale expanded from 4348 and 5443 to 7041 and 8771, separately.

4. RESULT AND DISCUSSION

1. Wasteland was most extreme in LPA zones of Madurai around 1200 sq.km which exceptionally high look at other land use likewise land valuation was exceptionally low in all area in the year 2007. In 2019 wasteland decreases to 800 sq.km shares the land with settlement and agriculture land practices. Period 12 years most of the waste land wasconverted to industrial, urban, semi urban because of population growth also. Also some agriculture practices were increased during these periods. This shows urban sprawl continues and increases the land valuations, there are various constructions, Road development happened during this period. This cause real estate practices are developed with iconic growth.

2. With reference to 2019, the farming area covers the greatest region square kilometer with 877 Sq.km contrast with 535 sq.km in 2007. This shows the agriculture practices increases in 2019. This may due to various sources such increases in Vagai dam Water flow or canal irrigation sources. This shows good greenery compare to 2019. Agriculture land in some locations were converted to settlement have been recorded in the landuse mapping.

2. Built-up (settlement) area rapidly increase in LPA areas because core city development towards the population and migration other activities like construction and road development increased in 2019 compare to 2007. The area increased from 149 Sq.km to 222 Sq.km during this period. Which directly proportional to land valuation factors.

3. The forest area has not changed. The barrel land has diminished from 1200 Sq.km to 800 Sq.km. Water bodies has indicated a slight abatement to 220.441 Sq.km from 237 Sq.km.

4. The GLR is expanded from 1.4 to 19 inside 12 years. The GLR for B and C district expanded from 1241 and 2774 to 4000 and 5450.

5. The PMR is expanded from 10 to 119 inside 12 years. The PMR for B and C district expanded from 4348 and 5443 to 7041 and 8771

5. CONCLUSION

The level of increment of GLR and PMR esteems in 2007 differs from 49.10% to 92.63%. The expansion level of GLR and PMR esteems in 2019 changes from 24.05% to 91.59%.

The level of increment in GLR and PMR esteems means that the expansion in land esteems because of development of urbanization. The progressions in landuse landcover are demonstrated by the expansion in GLR and PMR. The development region has created indeed because of the development in populace. The populace development expands the need to develop and settlement territory.

Landuse Landcover study important factor to understand land valuation over the period in LPA areas for better planning and development. There are multiple opportunities towards economic growth of the city between this period were created. For this reason the study has taken to compare urban sprawl and land valuation.

REFERENCES

1. Aitkenhead M J, & Aalders I H 2011 Automating land cover mapping of Scotland using expert system and knowledge integration methods *Remote Sensing of Environment* **115**5 1285–1295.
2. Bigdeli B, Samadzadegan F, and Reinartz P 2013 A multiple SVM system for classification of hyperspectral remote sensing data*Journal of the Indian Society of Remote Sensing*1–14.
3. Jannet, Bencure C, Nitin K Tripathi , Hiroyuki Miyazaki , Sarawut Ninsawat and Sohee Minsun Kim2019Development of an Innovative Land Valuation Model (iLVM) for Mass Appraisal Application in Sub-Urban Areas Using AHP: An Integration of Theoretical andPractical Approaches, MDPI*Sustainability*.
4. Longley PA, Goodchil MF, Maguire, DJ and Rhind DW 2010 Geographic Information Systems and Science (3rd ed.). New York: John Wiley and Sons, Inc.
5. Malleswara Rao TC, Jai Sankar G, Roopesh and Kumar T 2012 A hierarchical hybrid svm method for classification of remotely sensed data*Journal of the Indian Society of Remote Sensing* **40**2 191–200.
6. Maulik U and Chakraborty D 2011 A self-trained ensemble with semisupervised SVM: an application to pixel classification of remote sensing imagery*Pattern Recognition* 443 615–623.
7. Mokhamad Surianto, Sulaeman Rahman Nidar, Sylviana Maya Damayanti 2019The analysis of factors affecting land value and testing the accuracy of land value zone: case study in Blora, Indonesia*International Journal of Sustainable Real Estate and Construction Economics* **1** 3.
8. Sang L, Zhang C, Yang J, Zhu D, and Yun W 2011 Simulation of land use spatial pattern of towns and villages based on CA–Markovmodel*Mathematical and Computer Modelling* **54** 938–943.
9. Seto K, Fragkias M, Guneralp B and Reilly M 2011 A metaanalysis of global urban land expansion*PloS ONE* **6** 8 e23777.
10. ShougengHu,Qiuming Cheng, LeWang and DeyiXu 2013Modeling land price distribution using multifractal IDW interpolation and fractal filtering method,*Landscape and Urban Planning* **110** 25-35.

Prediction of Gas Composition of a Fluidized Bed Gasifier in a Sugarcane Bagasse- Modeling

P Manikandan[1] and V Christus Jeya Singh [2]
[1] Department of Mechanical Engineering, Vins Christian College of Engineering, Nagercoil, Tamilnadu, 629180, India
[2] Department of Mechanical Engineering, St. Xavier's Catholic College of Engineering, Nagercoil, Tamilnadu, 629003, India

*Corresponding e-mail: manips53@gmail.com

Abstract. The assessment of the performance of various biomass fuel sources in gasifier is a significant zone of study for the execution of this innovation in different applications, pertinent to remote villages. This paper reveals the prediction of the composition of producer gas in a fluidized bed gasifier using Sugarcane Bagasse as feedstock where air to be used as gasifying agent. The model dependent on equilibrium constant and material balance has been created to assess the syngas compositions. The developed model has been used to predict the composition of constituent gases by varying equivalence ratio (ER) between 0.1 and 0.35 and keeping the moisture content (MC) constant. The outcomes show that the mole portion of CO from sugarcane bagasse is higher as compared to rice husk and coir pith and the composition of H_2 and CH_4 where absorbed very close to rice husk. Hence it recommended that the sugarcane bagasse and coir pith shall replace the rice husk and coir pith in fluidized bed gasifier for rural applications.

1. Introduction

India produces more than 40 million metric tons (MMT) of sugarcane bagasse waste and this raw material are used less percentage in the paper industry. Through this the sugarcane bagasse can also be used to produce electricity at low cost, and the utilization of the biomass resources as fuel shall reduce greenhouse emissions. Therefore, biomass could play a vital role for generation of clean energy by replacing fossil fuels [1].

Many of the large scale industries produce more air pollution, which deteriorates the environment, human health and causes damage of property and landscape. The air pollution is mainly generated by the burning of fossil fuels for different applications. The fossil fuel also causes climate change and global warming and hence we have to find out new and cleaner strategies and fuel especially from sustainable power sources [2]. The share of the fossil fuels among the world total energy demand is reported as 80% contribute where biomass resources only 10-15% [3]. India has a huge opportunity for energy breeding from the accessible 500 metric tons of biomass per year. The agricultural areas of India depend on accepted biomasses much as firewood, animal couch, agricultural and other related applications. It is likewise seen that the biomass imperativeness control may add to 33% of complete energy usage in India [4&5].

One of the thermochemical process measure the biomass gasification is to change over biomass into sustainable energizes. Moreover the gasification process to deliver energy for warmth and force, syngas and condensed energizes, for example, diesel and gas, elective powers [6]. The idea of united turn and power creation from biomass is principally valuable to trim developing based process industries and rural power generation in emerging countries [7&8].

Biomass gasification is one of the widely using methods for producing syngas from the biomass in this process pollution is lesser than the burning of biomass fuel. There are different kinds of biomass gasifiers are utilized in power creation applications. The gasification process is relatively difficult and presently lots of researchers have been motivated on modeling of gasifier based on the gasification reactions the design, simulation and optimization of performance of gasifier [7].

The fluidized bed gasification were also analyzed technology is preferred because of its different preferences which incorporate the variety of fuel used, great heat transfer rate, even and well-regulated temperature of bed, encouraging gas–solid mixing, etc [9].

The dynamic displaying of a bubbling fluidized bed biomass gasifier was studied and this model has been validated to measurements from transient tests in a small-scale fluidized bed biomass gasifier. The results were compared to experimental results of a pilot-scale (FBG) and previous literature models. The transient response model is studied different working conditions, for example, biomass feed rate, equivalence ratio (ER), ambient temperature, and primary char amount in the bed [10].

An empirical and non- equilibrium model was developed in a fluidized bed biomass gasifier and the model predict the performance of gasifier and are to be compared n a 80kWh bubbling fluidized bed air biomass gasifier experimental results. A chemical balance model created by Pio (2020) overestimated the concentrations of H_2 and CO and underestimated CH. However, the efficiency was estimated accurately [11].

The synthetic balance model (CEM) and limited substance harmony technique (RCEM) for bubbling fluidized bed gasifier (BFBG) were developed for almond shell gasification using Aspen Plus and the estimated values were compared to the experimental results. In this modeling the increased cold gas efficiency has the higher gasification temperature of the process and to improve in production of H_2 with the increase in steam to biomass (S/B) ratio and increase in biomass moisture content of the biomass [12].

NASA has developed a simulation model with non-stoichiometric balance dependent on minimization of Gibbs free energy procedure. The model results showed that in a specific equivalence proportion as the gasification temperature increments, the mole fraction CO raises and a reverse trend was absorbed for CO_2. In a coal and biomass mixture, the percentage of coal increases at a specific equivalence ratio the mole portion of CO increments, with the expansion in gasification temperature and a reverse trend is reported in experimental work for CO_2. In a fuel mixture the increase in coal percentage at a specific equivalence ratio reduces the mole concentration of H_2 for all gasification temperature [13].

The gasifying agent was picked either air or steam based on the application. The gas composition and calorific values were predicted using a thermodynamic equilibrium model, depend upon the impact of gasification temperature, Oxygen to biomass proportion, and Steam to biomass proportion. The developed model was solved by commercial MATLAB software using 'Fsolve' function. The study carried out in the state of Punjab in India, using four agricultural biomasses [14].

The gas composition and the higher and lower heating estimations of the syngas created dependent on balance based model was created in Aspen Plus software. The equilibrium model estimated the syngas species, the char and tar yield and the elemental energy balances based on the water-gas shift reaction (WGS) and methanation reaction of gasification process. The equilibrium model was used to examine two biomasses with various moisture content and various gasification conditions and the predicted values were compared with experimental data [15].

A mathematical stoichiometric thermodynamic equilibrium model developed was used to estimate the exhibition of biomass gasification measure about the syngas yield and composition. The thermodynamic balance models are exact and helpful instruments for the assessment and comparison of gasification process. The performance of gasification process was estimated and it was approved for steam gasification and air-steam gasification [16].

2. Modeling

The overall chemical formula of sugarcane bagasse based on one mole of carbon is written as $CH_{1.55}O_{0.77}N_{0.001}$. The prediction of the physical characteristics char gasification carried out based on Equilibrium modeling [17].

The proximate and ultimate analysis used to determine the composition of individual elements of the biomass materials considered in this study shown in Table 1.

Table 1. Proximate and Ultimate analysis of the selected Biomass Materials

Fuel	Proximate analysis (%)			Ultimate analysis (%)				
	Combustible	Moisture	Ash	C	H	O	N	S
Sugercane Bagasse	83.9	4.4	4	44.3	5.7	45.5	0.2	0.007
Rice Husk	70.53	9.95	19.5	49.1	3.79	46.4	0.63	0.09
Coir Pith	69.2	6.35	15.1	44	4.7	43.4	0.7	0

It is accepted that the all gasification responses follow the thermodynamic equilibrium conditions, and the major components of the producer gas are H_2, CO, CO_2, H_2O, CH_4, N_2, and char content. This modeling assumed that the ash to be inert in the gasification reactions. The overall gasification reaction with air can be composed as:

$$CH_mO_nN_o + m_w\,H_2O + m_a\,O_2 + 3.76\,m_a\,N_2 = x_1\,H_2 + x_2\,CO + x_3\,CO_2 + x_4\,H_2O + x_5\,CH_4 + x_6\,N_2 \quad (1)$$

Where m, n, and o, are the mole number of Hydrogen, Oxygen and Nitrogen present in the biomass per one mole of biomass carbon, m_w is the moisture content present in the biomass and m_a is the mole of air supplied. In the above general equation, the moles of six unknown species of syngas are represented as x_1, x_2, x_3, x_4, x_5, and x_6. An equilibrium condition the mole numbers of reactants are equal to the mole numbers of the products. In order to estimate the six species concentrations, six equations are developed. The elemental balancing of Carbon, Hydrogen, Oxygen and Nitrogen in the general equation obtained the four linear equation and other two nonlinear equations are attained from the equilibrium constants [18].

From eqn. (1), the mole balancing for C, H, O and N are:

Carbon balance

$$1= x_2 + x_3 + x_5 \quad (2)$$

Hydrogen balance

$$m + 2m_w = 2x_1 + 2x_4 + 4x_5 \quad (3)$$

Oxygen balance

$$n + m_w + 2m_a = x_2 + 2x_3 + x_4 \quad (4)$$

Nitrogen balance

$$o + 7.52m_a = 2x_6 \quad (5)$$

In this model all chemical reactions are assumed to be in equilibrium condition, and the following reactions are playing vital role in the gasification process.

Boudouard reaction:	$C + CO_2 = 2CO$	(+ 172.6 kJ / mol)	(6)
Steam gasification:	$C + H_2O = CO + H_2$	(+ 131.4 kJ / mol)	(7)
Water gas shift reaction:	$CO + H_2O = CO_2 + H_2$	(- 41.2 kJ / mol)	(8)
Methanation reaction:	$C + 2H_2 = CH_4$	(- 74.9 kJ / mol)	(9)

It is assumed that all these chemical reactions are obey the perfect gas characteristics and the gasification is held at atmospheric condition. The following equations have no temperature component.

The water–gas shift reaction of equilibrium constant is given by

$$K_1 = \frac{p_{CO_2}\, p_{H_2}}{p_{CO} p_{H_2}} = \frac{x_1 x_3}{x_2 x_4} \quad (10)$$

The methane formation of equilibrium constant is given by

$$K_2 = \frac{p_{CH_4}}{p_{H_2}} = \frac{x_5}{x_1^2} \quad (11)$$

The values of above constants K_1 and K_2 are also calculated by the Gibbs free energy

$$K_1 = \exp\left(-\left(G^0_{T,H_2} + G^0_{T,CO_2} - G^0_{T,CO} - G^0_{T,H_2O}\right)/R_aT\right) \quad (12)$$

$$K_2 = \exp\left(-\left(G^0_{T,CH_4} - 2G^0_{T,H_2}\right)/R_aT\right) \quad (13)$$

Where, R_a is the general gas constant and G_T^0 is the standard Gibbs function at gasification temperature, T (K) of the gas species.

To evaluate the gas concentrations of the six unknowns, initially calculate the values of K_1 and K_2 at the particular temperature, and the values of K_1 and K_2 values are substituted in eqn. (10) and (11). A program was developed in SCILAB an open source software using the tool "Fsolve" to solve the six equations (2), (3), (4), (5) (10), and (11).

3. Results and discussion

The equilibrium model is used to predict the production of producer gas in a fluidized bed gasifier from the sugarcane bagasse. The impact of equivalence ratio on the gasification at temperature of 850°C and density of biomass is 194 kg/m^3 has been the choice of equivalence ratio was kept between 0.1 and 0.35 for this study based on the previous literature. The variation of the syngas concentrations of H_2, CO, CO_2, CH_4 and N_2 on dry basics are studied and reported shown in Figs. 1- 5. It is seen that expansion in ER reduces the concentration of H_2, CO and CH_4 percentage drops while CO_2 and N_2 increased.

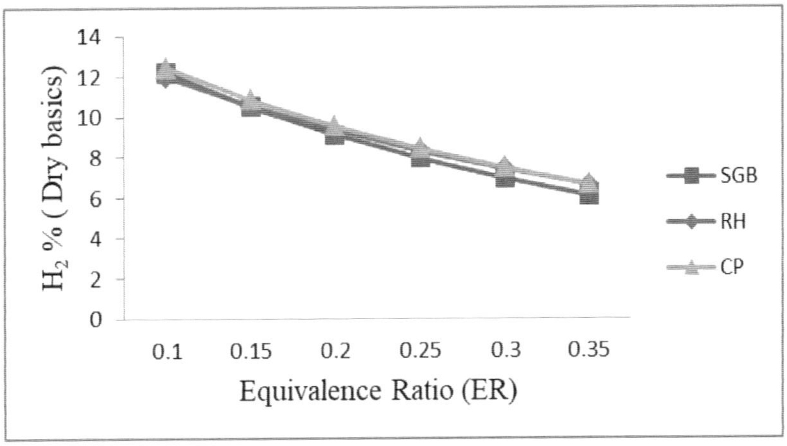

Figure 1. Variation of species concentration of H_2 with equivalence ratio

A common trend of decrease in the concentration of H_2 in producer gas for increase of equivalence ratio has been observed for all the biomasses (fig.1). The H_2 yield is maximum of around 12% when ER kept at 0.1 for all biomasses. This due to that the increase in ER beyond the value of 0.1, reduces the rate of reactions of steam gasification and water gas shift reactions.

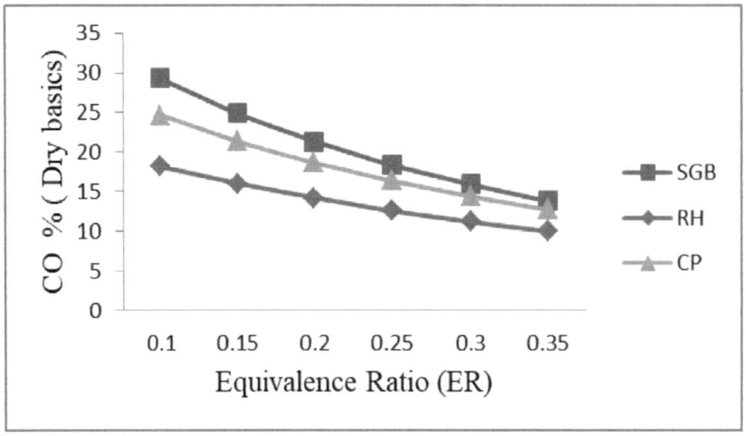

Figure 2. Variation of species concentration of CO with equivalence ratio

The analysis shows that the increase in equivalence ratio reduces the concentration of CO in product gas (Figure.2). All the three biomass materials, Sugarcane Bagasse, Rice Husk and Coir Pith have been reported with the same trend, when the ER was varied between 0.1 and 0.35. The CO yield at an equivalence ratio of 0.1 is observed with the maximum value of 29% for sugarcane bagasse as compared to the other two biomasses. This is due to the fixed carbon species in sugarcane bagasse is more than rice husk and coir pith.

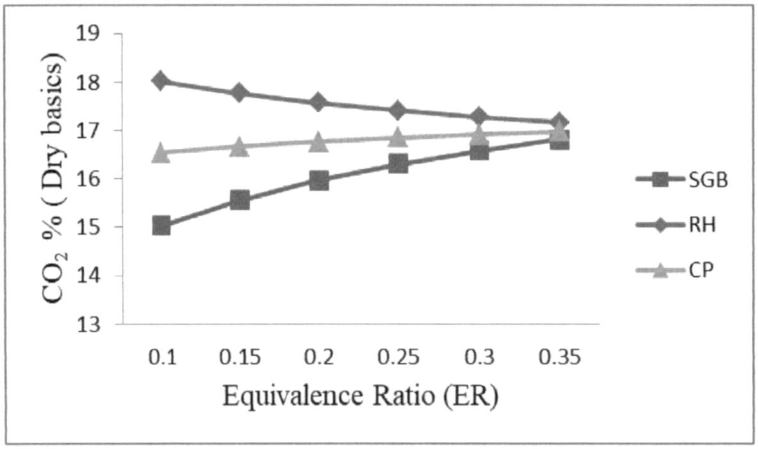

Figure 3. Variation of species concentration of CO_2 with equivalence ratio

In Figure. 3, In this prediction the equivalence ratio is increased the CO_2 yield is increased in Sugarcane Bagasse, Rice Husk and Coir Pith Estimated CO_2 yield is decreased. The CO_2 yield is 16.5% when ER is 0.35 in this prediction for Sugarcane Bagasse is maximum yield of CO_2. The variation of the species concentration of CO_2 with equivalence ratio for three biomasses is shown in Fig.3. An increasing trend for sugarcane bagasse and coir pith and a reducing trend were observed for rice husk when the equivalence ratio was varied 0.1 to 0.35. The fixed carbon content in sugarcane bagasse also promotes the production of CO_2 when the equivalence ratio is increased. However the CO_2 yield is obtained close to 15% at equivalence ratio is 0.1.

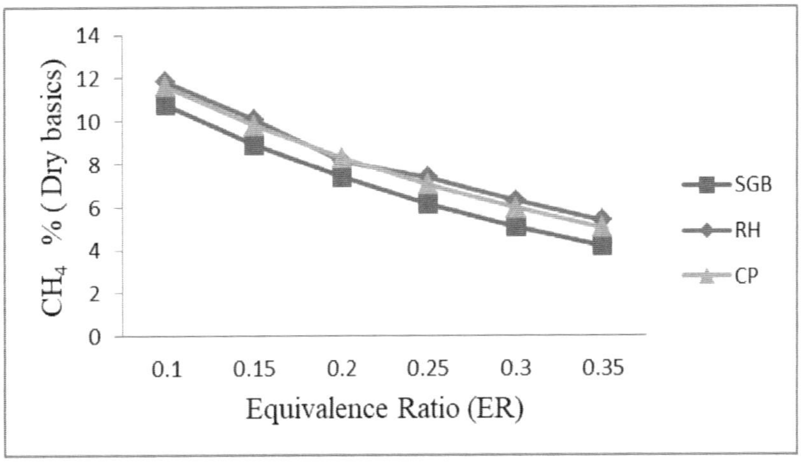

Figure 4. Variation of species concentration of CH_4 with Equivalence ratio

The variation of concentration of CH_4 obtained for different equivalence ratios, for Sugarcane bagasse, Rice husk and Coir pith are presented in Figure.4. When the equivalence ratio is increased, the CH_4 yield is decreased, and the CH_4 yield for sugarcane bagasse is 11.5% when ER is kept at 0.1. It is observed from the study that the increase in equivalence ratio affects the methanation reaction.

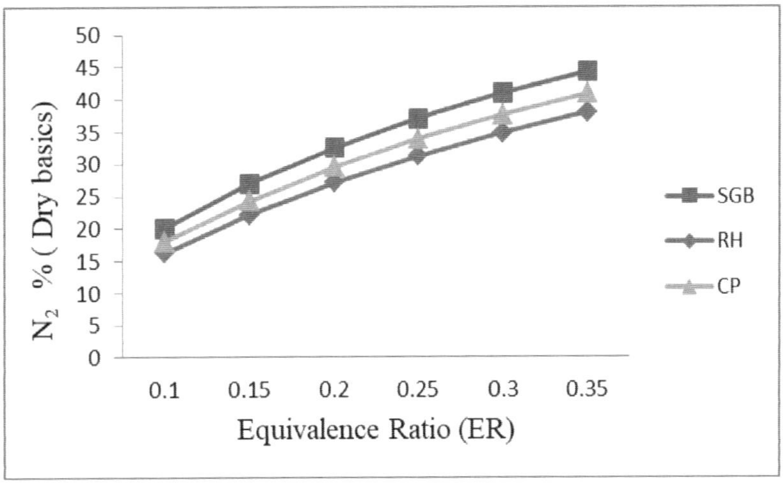

Figure 5. Variation of species concentration of N_2 with equivalence ratio

The inert gas Nitrogen yield is increased when increasing the equivalence ratio for all the three biomass materials and this has been shown in Figure.5. This is due to the content of N_2 in air supplied is increased when the equivalence ratio is increased. However, the concentration of N_2 was reported as minimum for Sugarcane bagasse, Rice husk and Coir pith.

4. Conclusion

In this study the prediction of gas composition sugarcane bagasse, coir pith and rice husk using equilibrium model in a fluidized bed gasifier from has been analyzed. The gas compositions of the producer gas have been studied by keeping the various equivalence between 0.1 and 0.35. The study showed that the maximum production of H_2, CO and CH_4 where obtained at the minimum equivalence ratio of 0.1. On the other hand the composition of CO_2 and N_2 were increased with increase in equivalence ratio. The maximum combustible gas was obtained at various equivalence ratios for sugarcane bagasse as compared to rice husk and coir pith. Hence it is suggested that the sugarcane bagasse can replace the other two biomasses where it is available abundantly.

References

[1] Kalpana Bisht 2016 Bagasse Power an Untapped Potential in India -A Review *International Journal of Engineering Sciences & Research Technology* 2277-9655.
[2] Chanchal Loha 2011 Performance of fluidized bed steam gasification of biomass – Modeling and experiment *Energy Conversion and Management* **52** 1583-1588.
[3] Faaij A 2006 Modern biomass conversion technologies *Mitigation and Adaptation Strategies for Global Change* **11** 343-375.
[4] Sreejith C C 2013 Performance prediction of steam gasification of wood using an ASPEN PLUS thermodynamic equilibrium model *International Journal of Sustainable Energy* **33** 416-434.
[5] Kumar A 2009 Thermochemical biomass gasification: A Review of the Current Status of the Technology *Energy* **2** 556-81.
[6] Report of Pacific Northwest National Laboratory (PNNL) & National Renewable Energy Laboratory (NREL) 2004.
[7] Son Yi 2006 A Study on measurement of the light tar content in the fuel gas produced from small-scale gasification and power generation systems *Proceedings of the 15th Annual Meeting of the Japan Institute of Energy* Japan August 3rd-4th.
[8] Thomas B R 1998, Hand Book of Biomass Downdraft Gasifier Engine System Golden Colorado *The Biomass Energy Foundation Press.*
[9] Cui H 2007 Fluidization of biomass particles: a review of experimental multiphase flow aspects *Chem Eng Sci* **62** 1 45–55.
[10] Suarez-Almeida M 2020 Modeling the transient response of a fluidized-bed biomass gasifier, *Fuel* 270 – 117226
[11] Pio D T 2020 Empirical and chemical equilibrium modelling for prediction of biomass gasification products in bubbling fluidized beds *Energy* 202 – 117654
[12] Mahmut Caner Acar 2019 Simulation of biomass gasification in a BFBG using chemical equilibrium model and restricted chemical equilibrium method *Biomass and Bioenergy* **125** 131–138
[13] Samar Das 2018 Thermodynamic optimization of coal-biomass co-gasification process by using non-stoichiometric equilibrium modeling *Materials Today Proceedings* **5** 23089–23098
[14] Vinita Shirke 2018 Equilibrium Model for Biomass Gasification: Study of effect of Biomass properties and Operating parameters *Materials Today Proceedings* **5** 22983–22992
[15] Gagliano A 2017 Development of an equilibrium-based model of gasification of biomass by Aspen Plus *Energy Procedia* **111** 1010 – 1019
[16] Joel George 2016 Stoichiometric Equilibrium Model based Assessment of Hydrogen Generation through Biomass Gasification *Procedia Technology* **25** 982 – 989
[17] Venkata Ramanan M 2008 Performance Prediction and Validation of Equilibrium Modeling for Gasification of Cashew Nut Shell Char *Brazilian Journal of Chemical Engineering* 585 – 601
[18] Chanchal Loha 2011 Thermodynamic analysis of hydrogen rich synthetic gas generation from fluidized bed gasification of rice husk *Energy* **36** 4063-4071

IOP Conf. Series: Materials Science and Engineering **983** (2020) 012014 doi:10.1088/1757-899X/983/1/012014

Segmentation of cervix using minimum spanning superpixel tree detector

Sheela Shiney TS [1*], Albert Jerome S[1*], Jemila Rose[2*]

[1]Assistant Professor, Department of Computer Science and Engineering , St. Xavier's Catholic College of Engineering, Chunkankadai, Tamil Nadu, India.

[1]Associate Professor, Department of Biomedical Engineering, Noorul Islam Centre for Higher Education, Kumaracoil ,Tamil Nadu, India.

[2]Associate Professor, Department of Computer Science and Engineering , St. Xavier's Catholic College of Engineering, Chunkankadai ,Tamil Nadu, India.

*Corresponding Author sheela@sxcce.edu.in.com , albertjerome@niuniv.com, jemila.rose@gmail.com

Abstract: Cervical cancer occurs in cervix region of women which is life threatening and is a key research in the field of medical diagnosis. Cervical cancer can be prevented if precancerous changes is diagnosed earlier and cured properly. But finding cancer at the earlier stage is a tedious process. Cervical cancer screening is crucial since utmost of the screening procedures are invasive in nature. The objective is to automatically extract the area where in the cervical cancer starts to occur. In order to diagnose cervical cancer and distinguish the malignant and benign tissues, a cytology image obtained by cervicographic or colposcopy device is used. A robust segmentation algorithm is proposed that uses superpixel for the image segmentation method. This method is regularized by the combination of structural properties such as color, texture, and spatial data present in the superpixel graph.

1. Introduction

Cervical cancer is the fourth cancer worldwide which develops in woman's cervix. Nearly all cervical cancer cases get infected with high-risk human papillomaviruses (HPV), an extremely common virus transmitted through sexual contact. The pre-cancerous changes may lead to cervical cancer and only few women gets affected by cancer. It may take nearly ten years for the pre-cancerous changes to change to cancer but in worst cases it may occur within a year. But the symptoms of cervical cancer can be seen only in advanced stage and leads to death. Getting a regular screening or pelvic examination can highly reduce the risk of pre-cancers turning into cancer. The screening procedure involves the great deals of skill which is associated in the screening procedure. It can be done only by a trained and knowledgeable cytotechnologist or a pathologist. Hence a novel method is required to overwhelm the death caused by cervical cancer in women by diagnosing at the initial stages.

2. Literature survey

Alan Conrad Bovik *et al.,* suggested a representation to approximate the cell contour as a set of sparse contour points [1]. For the partial segmentation of overlapping cells from images of cervical smear contour points are extracted. To enhance the true edges a gradient decomposition- method is applied which is based on edge enhancement. Furthermore, to identify the feeble contour points a contour search method is implemented. Finally, to remove the exact cell boundary Gradient Vector Flow Snake model is performed.

The cervical cancer is diagnosed earlier by the automatic segmentation of nuclei from the microscopic images proposed by Brette L. Luck *et al.,* [2]. The model is implemented by Gaussian Markov random fields which is combined with anisotropic median-diffusion filter. Finally classification is performed using Bayesian framework. The algorithm is implemented using cervical dataset images and it has been observed that it predicts 90% of manual-segmented nuclei having around six false positives per image.

An automated segmentation of major lesions is proposed by Changick Kim *et al.,* so that the cervical cancer is diagnosed in the early stages [3]. The main aim of segmentation is to find the location for a biopsy automatically. The aceto-white region is segmented by using a non-convex optimization algorithm.

An automatic cell detection process is performed to overcome the overlapped nuclei is proposed by Chanho Jung *et al.,* [4]. The distance transform is applied to overlapped nuclei. In this proposed algorithm the distance transform displays the area to be extracted. Next the parametric expectation-maximization is done in order to acquire knowledge about the area. To determine number of overlapping nuclei cluster validation is accomplished.

Christophoros Nikou *et al.,* suggested a fully automated approach for detecting cell nuclei in Pap smear images [5]. Using morphological analysis the contour of the cell nuclei are detected and they are refined, which includes a priori knowledge about the circumference of each nucleus. Thirty eight cytological images was evaluated by using the proposed method recognized around 5617 cells are squamous epithelial cells.

P. Dollar *et al.*, recommended that accurate segmentation of cervical cells in Pap smear images is significant to detect the precursor changes in cervix [6]. This study introduced a learning-based method to fragment distinct cell in Pap smear data in order to overcome the overlapping cell nuclei. To demonstrate the segmentation accuracy of our proposed work an evaluation metrics is performed using 2 dissimilar datasets.

Jacob Goldberger *et al.,* discussed a segmentation scheme which segments automatically of main lesions is detected in primary phases of cervix region [7]. To identify the position of cervix for a biopsy segmentation is done automatically. By using a non-convex optimization approach the aceto-white region of cervix, is first segmented. Within the aceto-white region, by texture analysis further abnormal features are classified. The extracted features evidently discriminates the normal and abnormal regions using fuzzy c-means algorithm.

To segment nuclear and cytoplasmic contours, a semiautomatic PC-based cellular image analysis system was developed by Marina E. Plissiti *et al.,*[8]. The first experiment result shows the cross-validation results of the average accuracies of 97.16% and 98.83%, respectively for four-cluster and two-cluster classifiers. The second experiment result shows the accuracy of 96.12% for four-cluster and 98.61% for two-cluster classifiers.

Micheal Hocket *et al.,* suggested that visual information management image coding for effective communication [9]. To develop tools for analysing the image content is a perplexing mission for vision researchers and engineers In this paper using color and texture features, the properties of three tissue types are modelled.

Philip King *et al.,* suggested an alternative screening tool named Cervicography that uses color cervix images which has been widely used [10]. An automated registration of Cervigram TM images are introduced in this technique. In this work a segmentation-based registration is defined and a combination of clustering- and active contour-based methodology is utilized.

Wei Wang *et al.,* proposed a distance-guided learning technique implanted in a multiple classifier system (MCS) for segmentation of tissue from the optical images of the uterine cervix [11]. In order to compute the decision boundary shapes similarity between a pair of statistical classifiers distance based metric has been proposed.

Sadao Isotani *et al.,* proposed a system for the determination of image density and morphometry for the analysis of malignant nuclei [12]. In this work Nuclei Demarcation is carried out for regions which are dark, the segmentation is performed by using the area threshold and finally the regions are validated to optimize the functionality of SPCIM.

A novel technique is proposed by V. Pallavi *et al.,* [13] to grade the cancer severity. The Lugol's iodine image is segmented to classify the abnormal tissues to diagnose the abnormal tissues accurately. This information is further used for grading the cancer.

Wei Wang *et al.,* proposed that regular screening of cervix can detect pre cancer changes earlier [14]. In this work a lightweight yet elegant computer supported automatic method is proposed.

Zhiyun Xue *et al.,* presented learning class-specific boundaries a method for extraction and segmentation of a class-specific object [15] automatically. Using a Markov random field (MRF) the watershed segmentation map of the input image is modelled which indicates whether the region is cancerous tissue or not.

It is important to develop tools to improve the reliability of cancer diagnosis for automated cancer diagnosis. The cancer diagnosis tool consists of four main computational steps: pre-processing, segmentation, feature extraction and classification. The cytology image obtained by cervicographic or colposcopy device is subjected to noise during staining process. The pre-processing step contains noise removal phase to remove the noise and to enrich the image quality to classify the objects and segmentation phase for extracting nucleus/cell information and abnormal regions. The feature extraction and diagnosis phase depends on the successful rate of segmentation results.

3. PROPOSED SYSTEM

The proposed architecture of the framework is exemplified in Figure. 1 .The main objective of this work is to segment the tumour region from cervicography image.

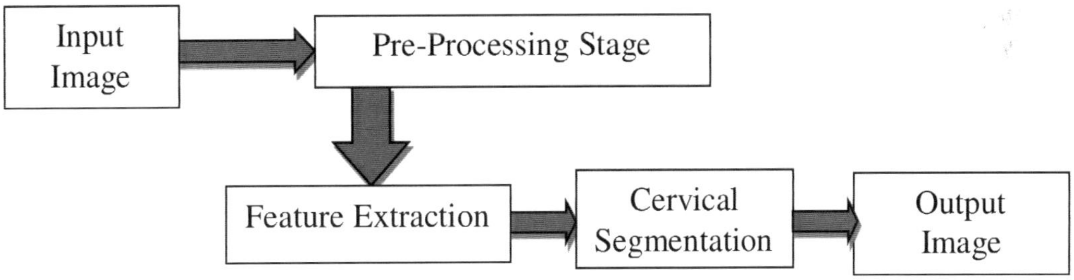

Figure. 1 Block Diagram

The noise information present in the input image is pre-processed by suitable pre-processing algorithm. Pre-processing is performed by using contrast enhancement method and cervical boundary growth. Then the image is being processed in feature extraction model. The feature extraction model consists of Illumination Layer, Reflectance layer, Texture layer. Finally the Minimum Spanning Superpixel Tree Detector is used to segment the cervical image.

An effective Simple linear iterative clustering (SLIC) algorithm is implemented to segment the cervical image into a group of superpixels.

a. Feature Extraction
The Feature Extraction comprises of illumination layer, reflectance layer and Texture Layer.

Illumination Layer and Reflectance layer

The input image is expressed mathematically as illustrated in equation 1.where L1 denotes illumination layer and L2 denotes reflectance layer

$$I = L_1 + L_2 \quad (1)$$

Texture Layer

In order to analyse the texture features Gabor transform is mathematically expressed in the equation.

$$T_\varphi(b,\theta,a) = c_\varphi^{-1/2} a^{-1} \int \varphi^*\left(a^{-1} r_{-\theta}(x-b)\right) I(x)\, d^2x \qquad (2)$$

Where, I is the input image, $A = diag\left[\eta^{-1/2}, 1\right]$ represents 2x2 matrix in which non-diagonal values are zero, k0 designates a vector frequency. ψ, ψ*, a, and b are the two dimensional Gabor variables.

To attain clear cervical position in different angle, the maximum wavelet feedback for each pixel is calculated using the equation 3.

$$M_\varphi(a,b) = \max_\theta \; |T_\varphi(b,\theta,a)| \qquad (3)$$

Where θ. Ranges from 0 to 170

c. Superpixel based Segmentation

Group of pixels which share similar characteristics is termed as superpixel .superpixel carry more information than the pixels .The input image is partitioned into number of segments that constitute a set of pixels or superpixels. .The number of superpixels in input image can be generated by using minimum spatial distance among various pixels. For computing super pixel the total number of pixels in the image is required. Then all the superpixels have been merged, till the number of superpixels is equivalent to the number of groupings contained in image.

SLIC based Superpixel Segmentation

Based on the color and proximity of image the Simple Linear Iterative Clustering (SLIC) algorithm generates superpixels by grouping similar pixels in an image. Superpixel can be generated by using five dimensional [r g b x y] space where r g b specifies pixel color values and x y denotes the coordinate position of pixel. To cluster pixel in this five dimensional space a distance measure is introduced.

Minimum Spanning Superpixel Tree Detector

In this paper, in order to employ the SLIC algorithm take illumination and texture layer information, to divide the input image into a cluster of superpixels. The cluster centres are located at each pixels. In order to prevent positioning cluster centres at the edges move the cluster centres towards the deepest gradient position. Subsequently the clustering process is applied on each pixels to the nearby cluster center. The overall distance measure D is expressed in equations (4) to (7).

$$d_{rgb} = \sqrt{\left(R_{k_1} - R_{k_2}\right)^2 + \left(G_{k_1} - G_{k_2}\right)^2 + \left(B_{k_1} - B_{k_2}\right)^2} \qquad (4)$$

$$d_t = \sqrt{\left(g_{k_1} - g_{k_2}\right)^2} \qquad (5)$$

$$d_s = \sqrt{\left(x_{k_1} - x_{k_2}\right)^2 + \left(y_{k_1} - y_{k_2}\right)^2} \qquad (6)$$

$$D = \sqrt{m_{rgb} d_{rgb}^2 + m_t d_t^2 m_s d_s^2} \qquad (7)$$

where R,G,B denote the red, green, and blue frequencies of the illumination layer, the pixel position, and the cluster center coordinate points respectively. Furthermore , d_{rgb}, d_t, and d_s represent

IOP Conf. Series: Materials Science and Engineering **983** (2020) 012014 doi:10.1088/1757-899X/983/1/012014

the illumination, texture, and spatial proximities, respectively, and m_{rgb}, m_t, and m_s represents the weights of the illumination distance, texture distance, and spatial distance, respectively.

4. Results and Discussion

a. Pre-processing

The input image is given in Figure. 2 and it is resized as shown in Figure. 3 and this image is converted into grayscale image as illustrated in Figure. 4. In order to enhance the contrast and to detect the boundary of cervix pre-processing step is applied. The local contrast enhancement is achieved by using Contrast Limited Adaptive Histogram Equalization (CLAHE) algorithm and is shown in Figure. 5. In addition, to drawbacks generated around camera aperture border. The wavelet transformation is applied to eliminate the artefacts and is shown in Figure. 6.

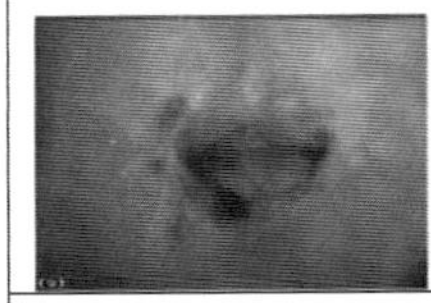	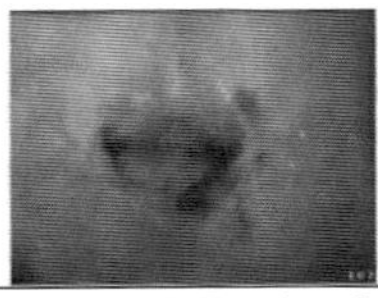	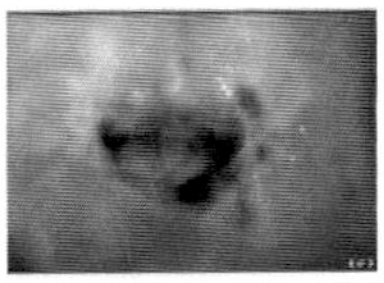	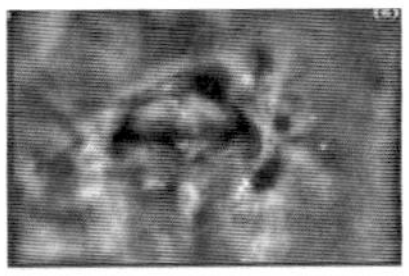	
Figure.2 Input Image	Figure.3 Resized Image	Figure. 4 Grey Scale Image	Figure.5 Contrast Enhanced Image	Figure.6 Cervical Boundary image

b. Feature Extraction

In order to extract the features, a probability-based method is applied so as to obtain illustration of the input image. From the input image, Illumination Layer is obtained and is given in Figure. 7. Reflectance Layer is also obtained and is shown in Figure. 8. The reflectance layer seems to be, illuminated wherein some portions are distributed unevenly. The Texture Layer given in Figure. 9 is computed from Cervical Boundary image.

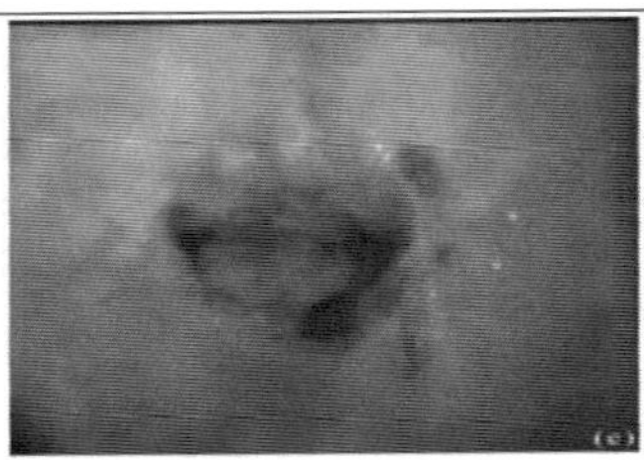	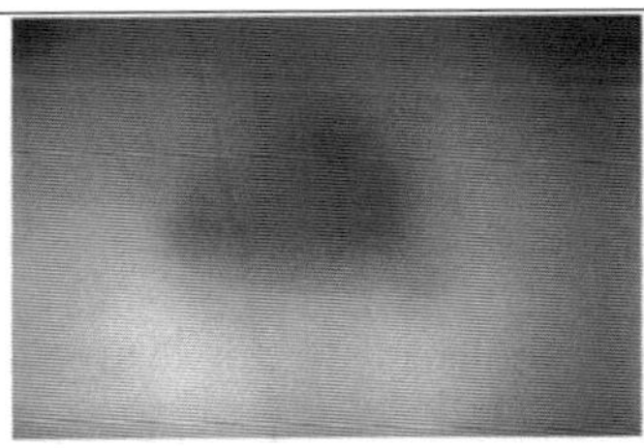	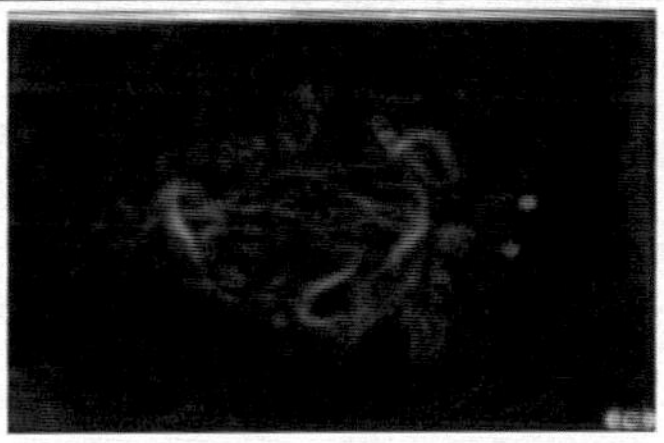
Figure 7. Illumination Layer	Figure 8. Reflectance Layer	Figure 9. Texture Layer

c. Segmentation

The Super Pixel Image is shown in Figure. 10 and the final segmented output is shown in Figure. 11.

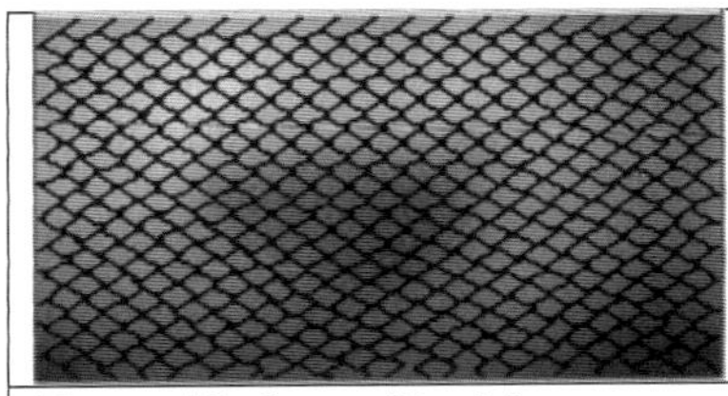	
Figure 10. Super Pixel Image	Figure 11. Segmented Image

5. Conclusion

Segmentation of cervix is a crucial task in the field of medical imaging. Although the image has artefacts and complex pathologies, existing methods have failed to appropriately segment poor contrast cervical image. Henceforth a superpixel based segmentation algorithm is introduced which operates on group of pixels rather than a single pixel. Furthermore a Minimum Spanning Tree (MSST) is implemented to evaluate each superpixels.

References

[1] Alan Conrat Bait, Barinova O, Kuzmishkina A, Vezhnvets A and Vezhnvets V 2007 Accurate Segmentation of Partially Overlapping Cervical Cells Based on Dynamic Sparse Contour Searching and GVF Snake Model in Proc. *Graphicon* 162–165.

[2] Brette S. Luck L, Belongie S, Malik J and Puzicha J 2002 An Image Model and Segmentation Algorithm for Reflectance Confocal Images of In Vivo Cervical Tissue *IEEE Trans. Pattern Anal. Mach. Intell* **24** 4 509–522.

[3] Changick Kim, Boykov Y and Funka-Lea G 2006 Graph cuts and efficient N-D image segmentation *Int. J. Comput. Vis* **70** 2 109–131.

[4] Chancho Jung G, Canny J 1986 Unsupervised Segmentation of Overlapped Nuclei Using Bayesian Classification *IEEE Trans. Pattern Anal. Mach. Intell* **8** 6 679–698.

[5] Cristopher Nikon PM, Gerbaldo D, Perino AR, Piccoli FJ, Montz and Captianio GL 1995 Automation Detection of Cell Nuclei Using Pap Smear *Obstet. Gynecol* **85** 1011–1016.

[6] Dollar P, Tu Z and Belongie S 2006 Accurate Cervical Cell Segmentation from Overlapping Clumps in Pap Smear Images in *Proc. Comput. Vis. Pattern Recognit (CVPR)* 1964–1971.

[7] Jacob Gold Berger B, Eluf-Neto J and Nascimento CM 2001 Segmentation and Classification of Cervix by Pattern and Texture *Semin. Oncol* 28 2 188–97.

[8] Marina.E.Plissiti C, Fei-Fei L and Perona P 2005 Semi-Automatic Segmentation and Classification of Pap Smear Cells in *Proc. Comput. Vis. Pattern Recognit. (CVPR)* 524–531.

[9] Micheal Hocket H, Felzenszwalb P 2001 Image Segmentation of Uterine Cervix Images for Indexing in PACS in Proc. *Comput. Vis. Pattern Recognit. (CVPR)* 56–62.

[10] Philip King J, Geraud T, Mangin J-F, Bloch I and Maitre H 1995 An Automated, Segmentation-Based, Rigid Registration System for Cervigram TM Images Utilizing Simple Clustering and Active Contour Techniques in Proc. *Int. Conf. Image Process* 548–551.

[11] Wei Wang M, Gordon S, Zimmerman G, Long R, Antani S, Jeronimo J and Greenspan H 2006 Automated and Interactive Lesion Detection and Segmentation in Uterine Cervix Images in Proc. *SPIE Med. Imag* **6144** 1549–1556.

[12] Sadao Isotani R, Greenspan H, Gordon S, Zimmerman G, Lotenberg S, Jeronimo J, Antani S and Long R 2009 A System For the Density of Image and Morphometric Analysis in Proc. *IEEE Trans. Med. Imag* 454–468.

[13] Pallavi V, Haris K, Efstratiadis S N, Maglaveras N and Katsaggelos A K 1998 Segmentation and Classification oh the Cervix by Pattern and Texture Analysis *IEEE Trans. Image Process* **7** 12 1684–1699.

[14] Wei Wang, Herrero R, Hildesheim A, Bratti C, Sherman M E, Hutchinson M, Morales J, Balmaceda I, Greenberg M D, Alfaro M, Burk R D, Wacholder S, Plummer M and Schiffman 2000 Automated and Interactive Lesion Detection and Segmentation in Uterine Cervix Images *J. Nat. Cancer Inst* **92** 6 464–474.

[15] Zhiyan Xine, Herrero R, Schiffman MH, Bratti C, Hildesheim A, Balmaceda I, Sherman ME, Greenberg M, Cardenas F, Gomez V, Helgesen K, Morales J, Hutchinson M, Mango L, Alfaro M, Potischman NW, Wacholder S, Swanson C and Briton LA 1997 Computer assisted Cervical cytological Nucleus *Pan Amer. J. Public Health* **1** 15 362–375.

IOP Conf. Series: Materials Science and Engineering **983** (2020) 012015 doi:10.1088/1757-899X/983/1/012015

Synthesis of Biodiesel from Waste Cooking Oil and Emission Characteristics of Its blends

Niraj S. Topare[1*], Kiran D. Patil[1], Satish V. Khedkar[2]

[1*]School of Chemical Engineering, Dr. Vishwanath Karad MIT World Peace University, Pune-411 038, India

[2]Department of Chemical Engineering, College of Engineering & Technology, Akola-444 104, India

E-mail: niraj.topare@mitwpu.edu.in, nirajtopare06@gmail.com

Abstract. In converting different vegetable oils for the respective esters, at a minimum reaction temperature and faster rate of the fatty acids, the catalyst plays a very important role which can be changed into fatty acid esters (FAME), also known as biodiesel by transesterification reaction with methanol. Homogenous and heterogeneous catalysts have been identified as catalysts and among these catalysts. Homogeneous catalysts were chosen because it was possible to generate more biodiesel at a comparatively faster rate. As a feedstock, soybean waste cooking oil (SWCO) was used in this research work. With the help of a conventional mechanical stirrer reactor, SWCO with methanol was heated at various temperatures. As a catalyst, sodium hydroxide (NaOH) and potassium hydroxide (KOH) is used. It evaluated the effect of various parameters such as reaction temperature (60^{0}C, 70^{0}C, and 80^{0}C), the reaction time of 2 hours, and loading of the catalyst (0.5 wt. %, 0.7 wt. %, and 1 wt. % KOH and NaOH). Results show that biodiesel produced from soybean waste cooking oil was within the recommended standards of biodiesel fuel. The transesterification reaction using KOH catalyst was more effective than the NaOH catalyst. The yield and conversion of biodiesel produced from SWCO by using 0.5 wt. % KOH catalysts at 60^{0}C are 93.2 % and 96.16 % respectively while; the yield and conversion of biodiesel produced from SWCO by using 0.5 wt. % NaOH catalysts at 60^{0}C are 91.35 % and 94.5 % respectively. Blends of 10, 20, 30, 50 and 100% by volume of biodiesel derived from soybean waste cooking oil and diesel fuel were prepared as B10, B20, B30, B50 and B100. Biodiesel blends have ASTM standards that are similar to diesel fuel, and the ORSAT apparatus has used to exhaust gas.

1. Introduction

After a rise in petroleum prices, biodiesel production is, due to its environmental benefits, a very new and technical field for researchers [1]. It is an efficient alternative fuel for diesel engines developed by the sterility reaction of various feedstock products such as vegetable oils or methanol animal fats. The commodity is called biodiesel or methyl ester, a replacement for non-toxic, sustainable, biodegradable diesel fuel. It lowers global warming emissions of petroleum diesel by biodiesel [2], for example carbon dioxide. The biomass diesel weights 11% oxygen and has no aromatic and sulphur [3,4]. The key challenge in our world, in these ten years, is the price of fossil diesel, which will one day be depleted. Find an alternative way to build a diesel substitute. For human beings, biodiesel is an urgent task [4]. Biodiesel has become a growing area of concern because of the rising populations and the depletion of fossil fuel in some countries because of the rise in oil prices and environmental concerns about pollution from the gas vehicles [5,6]. With oil prices increasing and gas supplies declining, an alternative fuel source has never been required. Biodiesel is an organic energy compound which can be produced by a transesterification reaction from different triglycerides (oils), alcohol and a (acid or basic) catalyst. This is chosen because the most commonly used source of decreased carbon chains are fatty acids. In this reaction, the oil is subjected straight to alcohol transesterification that results in a molecule of biodiesel [7,8].

Some sources that can be used as raw material for biodiesel production are inedible oil, animal fats and vegetable oil [9]. However, vegetable oils from a plant used because it has similar fuel qualities to diesel fuel, not even for higher viscosity and low oxidative stability, which must be met before they are turned into organic diesel. For example, pure plant oils, straight vegetable oils (SVO) and waste vegetable cooking oils (WVCO), are the most common use of vegetable oils in the reaction to transesterification [10]. Waste vegetable oils, amongst other processed or high-quality oils, are one of the best natural sources. It is easier to collect from other business sectors (supermarket oils) than other oils [11]. Through the use of these oils as raw material, we can therefore reduce the cost of biodiesel production. Prevention of environmental contamination and low operating costs [12] are the principal advantages of using waste cooking oils for biodiesel production.

Due to its high waste disposal costs, many people dispose of waste cooking oils directly to the environment, particularly in rural areas. These oils must be treatment before they are released from the atmosphere to avoid pollution [13]. It is important to use waste cooking oils efficiently in order to reduce the costs of production of biodiesel. This study was conducted with different homogeneous catalysts in the standard mechanical stirrer reactor to conduct bio-diesel output from the SWCO. SWCO contains high FFAs so that a transesterification of alkalis is done using methanol and (KOH & NaOH) catalysts. It is also used to determine the impact on the performance and exhaust emissions of diesel engines through biodiesel mixtures B10, B20, B30, B50 and B100 with diesel fuel.

2. Materials & Methods

2.1. Materials

Raw SWCO 10 litres from the local hotels were obtained. The KOH, NaOH and Methanol grades of analytical (AR) (99.5%) were used for testing. The heating mantle with the mechanic stirrer for heating the mixture in the reaction was used for a two-liter batch reactor equipped with a water-cooled reflux condenser, and the mixture was continuously stirred in response to the traditional method transesterification. Biodiesel in all quantities is miscible with petrol. Biodiesel is often combined with diesel very frequently. The blend is commonly called B10, B20 or B50. The blend of 10% biodiesel and 90% diesel is indicated by the B10. In terms of the Cetane number [14], biodiesel is similar to diesel. The ignition quality of a fuel is this property. The combustion heat, pour point, cloud, viscosity, and oxidative stable of other characteristics showing biodiesel which are widely related to diesel [15]. Biodiesel can be contrasted with diesel in terms of both chemical and physical properties as outlined under table 1.

Table 1. Comparison by ASTM standards of the traditional properties of biodiesel with diesel

Property	Biodiesel	Diesel
Composition	FAME (C12-C22)	HC(C10-C21)
Standard	ASTM D6751	ASTM D975
Specific gravity (g/ml)	0.88	0.85
Kinetic viscosity (mm^2/s) at 40^0C	1.9-6.0	1.9-4.1
Cetane Number	48-60	40-55
Cloud point (^0C)	-3 to 12	-15 to 5
Pour point (^oC)	-15 to 16	-35 to -15
Flash point (^0C)	100-170	60-80

2.2. Systems requirements (specifications)

2.2.1. Batch Reactor

The reaction was performed in a 2-liter batch reactor consisting of an overhead mechanical stirrer, with an impeller inserted from the central opening of the reactive vial, which was fitted with a heating mantle, to allow a 2-liter batch reactor to be properly fixed. On one side of the batch reactor it is connected in order to feed the raw material (SWCO, catalyst and methanol) and to extract samples from the thermometer (up to 120^0C) along with the other side stoppers. The batch reactor specification as shown in Table 2 below.

Table 2. Specification of batch reactor for transesterification reaction

Type of reactor	Three-necked batch reactor
Material of Construction	Glass
Capacity of batch reactor	2 liters
Diameter of batch reactor (D_r)	15.6 cm
Diameter of central neck of reactor (d_1)	4 cm
Diameter of impeller (d_2)	3.15 cm

2.3. Experimental Procedure

The transesterification reaction was carried out in the three-necked batch reactor having capacity 2 liters for homogenous catalyst (KOH and NaOH) concentration i.e. 0.5 wt%, 0.7 wt % and 1 wt % at different reaction time 2 hours and at different temperature i.e. 60 ^{0}C, 70 ^{0}C, 80^{0}C and at atmospheric pressure. The experimental set-up for biodiesel production is illustrated in Figure 1.

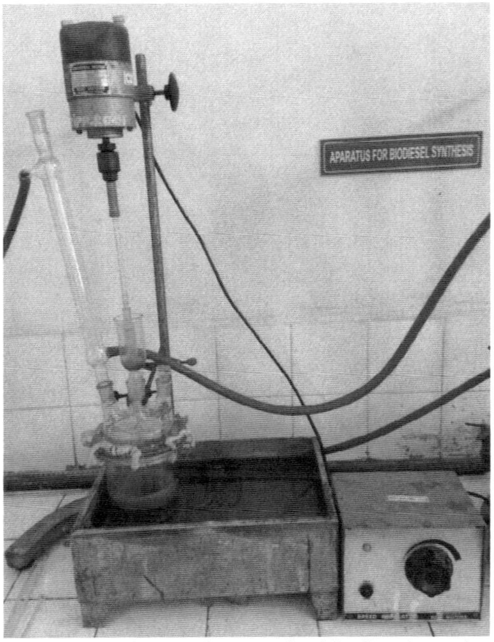

Figure 1. Experimental setup for biodiesel production

SWCO and methanol sample was measured carefully by measuring cylinder. The amount of catalysts was weighed and taken in a batch reactor equipped with a stirrer, thermometer, and condenser. The reactor containing the mixture was kept on the water bath. The mixture was heated up to different temperature 60 ^{0}C, 70 ^{0}C, and 80^{0}C with continuous agitation for 2 hours respectively. As the boiling point of methanol is 60^{0}C at this temperature it is evaporation, then it is condensed and refluxed back to the flask. When the reaction was finished, the outlet on the lower side of the reactor was drawn from the product and pour into the separating funnel which was mounted. Biodiesel, alcohol and soap were the upper layer. The bottom layer was glycerin, alcohol in abundance, catalyst, contaminants and traces of unreacted oil. The upper layer was purified: the alcohol removal was achieved by holding the mixture at a high temperature of 80^{0}C and the saponified substance was removed-with warm water washing. The process flow diagram for SWCO development of biodiesel is shown in Figure 2.

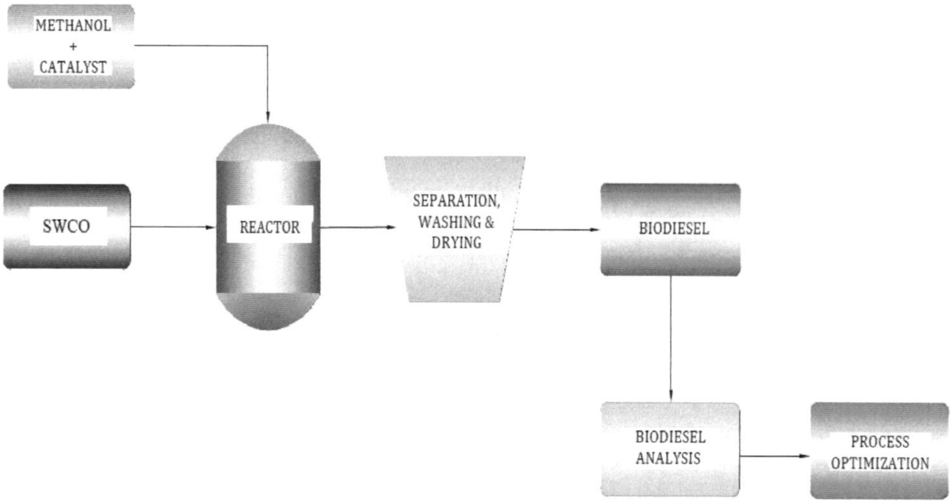

Figure 2. SWCO biodiesel development process flow diagram

3. Results and Discussions

The reaction was carried out for different homogeneous catalyst 0.5 wt%, 0.7 wt% and 1 wt % KOH and NaOH respectively at 60 ^{0}C, 70 ^{0}C, 80^{0}C. Figure-3, 4, 5 and 6, 7, 8 shows the effect of the amount of catalyst (KOH and NaOH) on the yield and conversion of SWCO respectively at 60 ^{0}C, 70 ^{0}C, 80^{0}C. The maximum yield was obtained for SWCO at 0.5wt% of KOH and NaOH catalyst respectively. It has also shown that the as amount of catalyst increases the percentage of yield and conversion decreases. Results show that the yield and conversion of biodiesel produced from SWCO by using 0.5 wt. % KOH catalysts are 93.2 % and 96.16 % respectively while; the yield and conversion of biodiesel produced from SWCO by using 0.5 wt. % NaOH catalysts are 91.35 % and 94.5 % respectively. It is possible to assume that the biodiesel production and transformation was greater than the NaOH catalyst with the use of the KOH catalyst. Acceptable operating conditions were met at 60^{0}C reaction temperature, a reaction time of 2 hours and a KOH concentration of 0.5 wt. %.

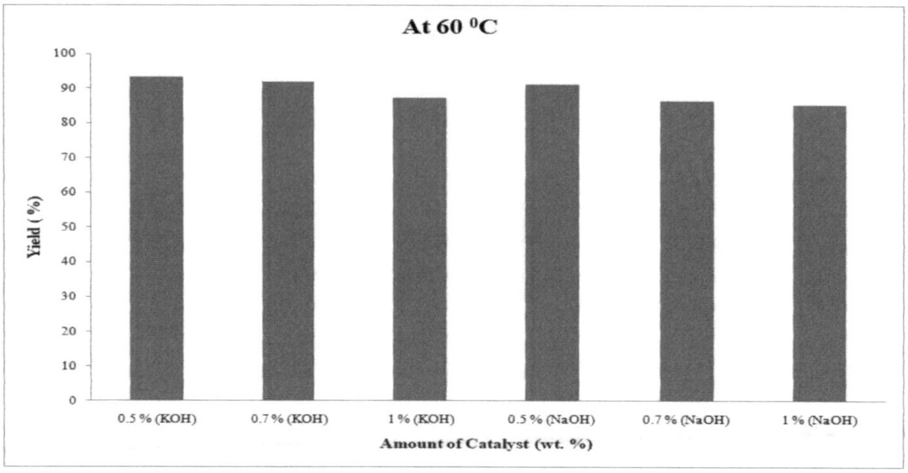

Figure 3. Yield (%) of biodiesel produced from SWCO by using KOH and NaOH catalyst at 60^{0}C

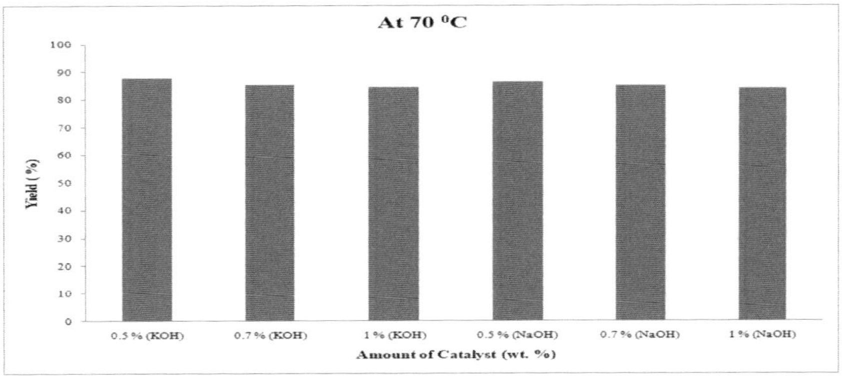

Figure 4. Yield (%) of biodiesel produced from SWCO by using KOH and NaOH catalyst at 70^0C

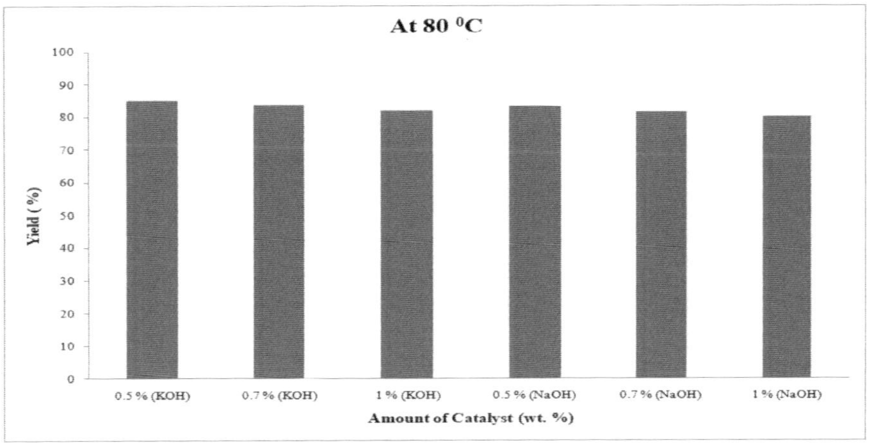

Figure 5. Yield (%) of biodiesel produced from SWCO by using KOH and NaOH catalyst at 80^0C

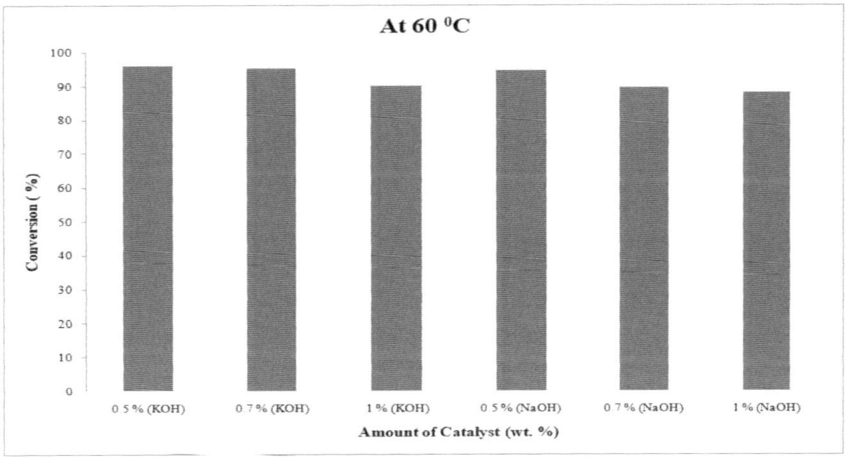

Figure 6. Conversion (%) of biodiesel produced from SWCO by using KOH and NaOH catalyst at 60^0C

IOP Conf. Series: Materials Science and Engineering **983** (2020) 012015 doi:10.1088/1757-899X/983/1/012015

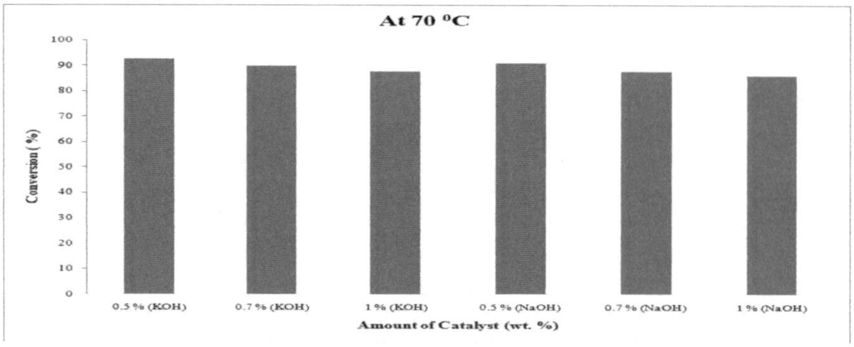

Figure 7. Conversion (%) of biodiesel produced from SWCO by using KOH and NaOH catalyst at 70^{0}C

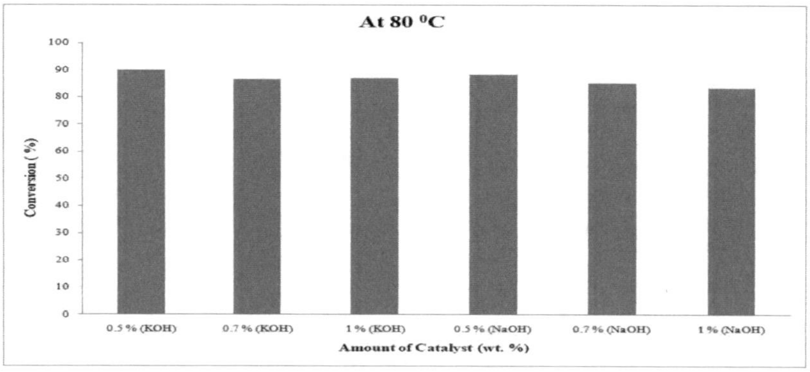

Figure 8. Conversion (%) of biodiesel produced from SWCO by using KOH and NaOH catalyst at 80^{0}C

The biodiesel produced from waste cooking soybean oil is used as an alternative fuel to operate diesel engines. Biodiesel blends are produced by the combination of biodiesel and oil diesel in acceptable proportions under appropriate circumstances. The tested engine provides a speed of 1500 RPM with a production ability of 3.73 kW for maximum utilization and engine specification as shown in Table 3. Biodiesel was used to measure the engine performance parameters with B10, B20, B30, B50 and B100 blends. The properties of biodiesel and its blends in comparison to diesel fuel are represented in Table 4. For measurements of exhaust gas concentrations, the ORSAT devices were used. It is found from the exhaust gas study that the CO_2 emissions of biodiesel are substantially lower than those of diesel as well as the mixture B10, B20, B30 and B50. Biodiesel has a much higher O_2 percentage than diesel and biodiesel CO is zero as shown in Table 5.

Table 3. Engine specification

Model	SF
Type	S
Serial No	11833
BHP	5
CEX No	12158
Method of starting	By using starting motor

Table 4. The properties of biodiesel and its blends in comparison to diesel fuel

Properties	Biodiesel (B100)	B10	B20	B30	B50	Diesel fuel
Density	0.940	0.65	0.78	0.82	0.82	0.84
Viscosity	3.46	1.53	2.22	2.85	3.1	3.28
Flash point	50^0C	52^0C	55^0C	56^0C	56^0C	55^0C
Fire Point	78^0C	79^0C	83^0C	85^0C	86 ^{0}C	80^0C

Table 5. ORSAT apparatus exhaust gas analysis

Sample gas (%)	Biodiesel	B10	B20	B30	B50	Diesel
O_2	18.2	8.5	9	9.2	10.4	5
CO_2	1.12	6	5	4	4	9
CO	0	1	1	1	1	1

4. Conclusion

Synthesis of biodiesel from SWCO using two homogeneous catalysts i.e. NaOH and KOH at percentages (0.5 wt %, 0.7 wt%, 1wt %) was performed. Results demonstrated that the required operative conditions at the reaction temperature of 60^0C, 2 reaction hours, and KOH concentrations at 0.5 wt. % were achieved. The yield and conversion obtained by using KOH catalyst was greater than the NaOH catalyst. It has also observed that less soap formation was obtained than with a KOH transesterification reaction. In addition, it is found that, due to the formation of soap, biodiesel derived from SWCO yield was reduced by increasing the NaOH and KOH catalyst concentration. Results also indicate that CO, CO_2 emissions for biodiesel were lower and that their mixtures were lower in comparison to diesel fuel.

References

[1] Singh A, Kumar G 2018 *J Biochem Tech* **9(1)** 17-27
[2] Abdelfatah M, Kiros Y Abu Elala R. 2017 *Petroleum & Petrochemical Engineering Journal*, **1(6)**, 1-7
[3] Xu W, Gao L, Wang S, Xiao G 2014 *Bio-resource Technology* **159** 286–291
[4] Thanh L T, Okitsu K, Boi L V, Maeda Y 2012 *Catalysts* **2** 191-222.
[5] Santos S, Gomes L N J, Puna J, Ferreira R Q, Bordado J 2019 *Energies* **12** 4670-4680
[6] Topare N S, Patil K D, Naik P, Sonawane A, Joshi P 2015 *Emerging Trends in Chemical Engineering* **2(1)** 1-8.
[7] Vasi V, Podrepsek G H, Knez Z, Leitge M 2020 *Catalysts* **10** 1-20.
[8] Thangaraj B, Solomon P, Muniyandi B, Ranganathan S, Lin L 2019 *Clean Energy* **3 (1)** 2–23.
[9] Hamamre Z A, Yamin J 2014 *Energy Conversion and Management* **79** 246-254.
[10] Diamantopoulos N, Panagiotaras D, Nikolopoulos D 2015 *Journal of Thermodynamics & Catalysis* **6 (1)** 1-8.
[11] Pandit PR, Fulekar MH 2019 *Renewable Energy* **136** 837-845.
[12] Mohamad M, Ngadi N, Lani NS 2015 *Advanced Materials Research* **1113** 518-522.
[13] Colombo K, Ender L, Barros AAC 2017 *Egyptian Journal of Petroleum* **26** 341–349.
[14] Shinde K, Kaliaguine S 2019 *ChemEngineering* **18 (3)** 1-10.
[15] Abed KA, Morsi AKE, Sayed MM, Shaib AAE, Gad MS 2018 *Egyptian Journal of Petroleum* **27** 985–989.

Synthesis of Polymer Nano-composite coatings as corrosion inhibitors: A quick review

Konica Sharma[1,*], M.S. Goyat[2], Piyush Vishwakarma[1]

[1]Department of Physics, Lingaya's Vidyapeeth, Faridabad,121002, Haryana, India
[2]Department of Physics, School of Engineering, University of Petroleum, Energy Studies, Dehradun, 248007, Uttarakhand, India

[*]Corresponding author: *E-mail address:* konicabhardwaj@gmail.com

Abstract: Corrosion- a big worldwide issue that destroys metal as its interaction takes place with the atmosphere. In recent years, polymer Nanocomposite coatings emerge as eminent materials for corrosion resistance. Coating means to deploy a layer over the metal interface that provides a shield to metal against corrosion. Polymer Nanocomposites are economic, Eco friendly, exhibits great mechanical strength, self-healing property, durability, chemical inhibition, and thermal stability that is the reason they have been widely used in various fields like an oil field, marine, defense, engineering industry and health industry too as corrosion inhibitors. In this review, the application of various polymer Nano-composites materials for corrosion inhibition with synthesis technique through reported materials are explored also we are identifying challenges that occur for the development of such Nanostructure coatings and overcoming that challenge for the future aspect of this field.

Keywords: Organic-inorganic polymers, nanoparticles, composite coatings, corrosion inhibitors, metal surface, polymer Nanocomposite coatings.

1. Introduction: nanotechnology fulfilling enormous purposes of research and manufacturing today. Nanoparticles are those particles whose size lies between 1 to 100 nm. Now they have been used in different fields for example–health and medicines[1]–[3], food[4], [5], agriculture[6], [7], engineering[8]–[11] and this all because of their great performance physically and chemically like robust mechanical strength, optically and chemically ready to react[12]–[14] and apart from all this one significant application they possess is to create inhibition over metals to safeguard them from corrosion in various atmospheric. conditions[15]–[17]. Metals-material with economical synthesis, extensive mechanical robustness having different applications like shipbuilding, drilling equipment, infrastructure, pipelines, etc. But this is just a positive side, all these applications of metals can be ruined by one process only that is corrosion. Corrosion is the chemical and electrochemical interaction of metals with the environment. Hence, the transformation of metal takes place into stable chemical forms like oxides, hydroxides, or sulfides. Moreover, Process changes or effects the structure and characteristics of various metals. Then there is no going back which means you cannot reverse the process, it ruined or destroy the whole material and it is very much costly. All these things make this one of the biggest issues around the globe. Corrosion cost is about 4% of gross national product (GNP)[18]–[20]. Physical and chemical properties; that is humidity, temperature, air, variation in pH, and electrolytes over the metal interface are major factors behind corrosion[21], [22]. Such conditions cannot be stopped completely but can be control or inhibited via various techniques like coating over metal interface, changing pH and potential through cathodic-anodic process[23]. A well known and important method is metal inhibition or metal. coating. Depositing or creating a safe coating over metal interface many inhibitors are used. Among all nano-composite materials is the best one because of their unique characteristics as compare to macroscopic materials. Moreover, they are environmentally friendly too.

2. Nano-composite coatings: In recent years nano-composite materials emerge as novel and better alternate serving various purposes of research and manufacturing. Nano-composites consist of heterogeneous films that have either chemically diverse phases or one phase isolated or embedded into a second phase. Phases in which one will be in the crystalline phase and the other in the amorphous phase or two distinct crystalline phases. Generally, these coatings are one host particle and another one embedded homogeneously in it with one or both in the nano-scale range. Nano-composite coatings possess unique characteristics that made them useful and striking for various engineering purposes like high-speed machining, tooling, optical devices, magnetic storage tools, moreover in automation, aerospace, and biomedical area among different other materials and that is all due to their unique mechanical, electronic, magnetic, optical characteristics because of size. This makes them a dynamic rising field, which exploits different methods and features of nanotechnology. So that one can create models and synthesize various interface films with extensive results and improved strength in less expenditure. Further, in these coatings there is space to realize novel and distinct functions through this one can get pioneering properties[24]–[30]. Among multiple coating applications, anti-corrosion is one that we are going to discuss in this review. In anti-corrosive nano-composite coating, we take inorganic nano-particles that possess strong and sustainable corrosion inhibition this is due to interaction among inorganic nano-material and host organic matrix, possessing less porous structure and less cracking ability of developed coating. From a wide variety of nano-composite materials, various techniques have been reported to synthesize nano-composite coatings having various complex functions like basic obstacle defense, rapid auto-response, and self-healing properties[31]–[37]. In this review, recent advances of nano-composites as corrosion inhibitors will be discussed.

2.1. Classification of nano-composite coatings:[25]

We can classify nano-composite on different criteria but if we classify them based on matrix materials, we can divide them into three types:

1. Ceramic matrix nanocomposites (CMNC)
2. Metal matrix nanocomposites (MMNC)
3. Polymer matrix nanocomposites (PMNC)

2.2. Polymer matrix nano-composite coating: polymer nanocomposites (PNCs) have matrix elements as polymer and dispersed material in the matrix is nano-particle. For ages, polymers are part of our life. From naturally existing polymers like Tar, tortoise shells, horns, etc, to manmade polymers like polyolefin, epoxies, resins offer variety for industry, commerce, research in the entire world. Macromolecules are characterized through several recurrences of one or more elements combined with an inadequate amount to offer various characteristics. So, polymers do not have uniformity chemically but the multiple-element system. By origin, polymers can be divided into two parts natural and synthetic. Cellulose, starch, dextran, chitosan, etc are natural polymers .synthetic polymers further are of three types: thermoplastics (polyethylene, polyvinyl chloride, etc), thermosets (melamine, urea, formaldehyde, etc), and Elastomer (natural rubber). Polymers possess various applications in automobiles, glues, medicines, ships, etc. This is all because of their eye-catching characteristics economic, environment-friendly, strength, stability, and many more moreover, that makes them suitable for corrosion inhibition[38], [39]. If we add fillers like ceramics, metals, etc researchers can develop numerous kinds of materials having special characteristics and that is too at less cost. Taking an example of the addition of filler into thermoplastics like polypropylene their outcomes are very less in cost than manufacturing plastic. In the same manner, if we mix the number of polymers to get resin that too enhances the properties of original polymers. Polymer nano-composites offer a variety of fillers to augment functions and applications of basic polymers and conquer conventional blended characteristics. A nano-composite term used initially in the 1980s which means a multi-constituent compound among those constituent major one is a polymer or its combination and minor part has a dimension of nanometer[40]–[43]. Polymer nano-composite is an inorganic-organic hybrid and elementary compound that can have well established industrial materials[44]. Polymer nanocomposites are the best replacement for conventional fillers of polymers, because of their nano-size and dispersion of fillers, which enhance their properties and make them unique and special to use in various applications. According to us, this is the right time to present a critical review on these polymer nanocomposite materials for one of their major application that is metal corrosion inhibition. We will try to give the knowledge for researchers, students, and industry making efforts in this field.

2.3. Synthesis techniques of Polymer nano-composites:

There are many techniques to synthesize polymer nano-composites from those mostly used and essential techniques are:

1. Intercalation [45]

2. In-situ polymerization[46]
3. Sol-gel process[47]
4. Melting extrusion[48]

Polymer nano-composites development is also influenced through molecular weight, in-organic constituents, size of constituents, and characteristics of inorganic constituents.

3. Polymer nanocomposites as corrosion inhibitor:

As to protect metal through polymer nano-composite techniques like wet-chemical reduction[49], photochemical reduction[50], laser mediated[51], hydrothermal[52],micro-wave assisted[53], etc are employed by various researchers. But if we see carefully we can divide all these methods into only two groups of techniques i.e. ex-situ and in-situ. In the ex-situ technique, we can encompass all methods in which nano-particle is developed outside the host matrix. Pre-developed nano-sized fillers are put in the host polymer matrix directly[54]. The advantages of this method are: permitting accurate control over size, shape, density, and particles, also it is approachable for industrial-scale synthesis. The basic issue with this method is the uniform scattering of nano-particles among matrix and maintaining stability against aggregation for a long period [55]. The next technique in-situ technique simply means the development of nano-fillers inside the polymer matrix. This is very basic and mostly can be done in single-pot. Rising nano-particles among matrix inhibit constituent agglomeration and then get sustaining spatial dispersion. That is why this technique is widely utilized by researchers. One challenge in this method is non-reacted constituents may affect the characteristics of the developed composite[56].

Mohamed Gamal et al[57]; Reported that they employ solution-dispersion method and heat ring-opening polymerization technique to synthesize a novel group of bisbenzylidene based polybenzoxazine (PBZ; CP-BZ or CH-BZ) composites having epoxidized soybean oil (E-SBO; 10 or 20 wt%) also E-SBO/bentonite (nano clay; 3 or 5 wt%) for application of corrosion-resistant coating over mild steel(MS). Their investigation on the chemical composition of CP-BZ and CH-BZ was done by Fourier transform infrared (FTIR) spectroscopy and ^{1}H and ^{13}C nuclear magnetic resonance spectroscopy. It was clear from differential scanning calorimetry (DSC) that the thermal polymerization temperature of CH-BZ before the experiment was extensively low that is $198^{0}C$ as compare to monomer 3-phenyl-3,4-dihydro-2*H*-benzoxazine that is $236^{0}C$. For the application of corrosion protection and to test coating for corrosion inhibition salt-spray and electrochemical techniques are used. A 20 weight% epoxy constituent mixed in polybenzoxazine epoxidized soybean oil (PBZ/E-SBO) gives rise to inhibition of corrosion greatly as compared to pristine PBZs. By adding 20 wt% E-SBO, and 3 wt% nano clay lower corrosion efficiency by 1 order of magnitude as compared to pristine alloy (CH-BZ) and two orders of magnitude when compared to uncoated mild steel (MS). They acquire a protection efficiency 98.16% that enlightened about the immense rise in inhibition characteristic of composite film for corrosion inhibitors. If we talk about the salt spray test, then it's applicable where a quick comparison of the expected and actual degree of corrosion resistance is required. Pure poly (CP-BZ) and pure poly (CH-BZ) shows scorching also the corroded area was much more than other coated film having E-SBO resin. That means coating had a severe chemical variation while exposing, heading towards failure. But in comparison with MS coated with POLY(CP-BZ)/E-SBO 10 or 20 wt% and poly (CH-BZ)/E-SBO (10 or 20 wt%) that reveals better resistance, even no corrosion dispersion with and just about crosscut, pointing towards excellent corrosion inhibition of film. On the rising value of content is of resin E-SBO between coatings that give rise to corrosion resistance, the reason behind this is the development of immensely crosslinked PBZ/E-SBO epoxy resin composite while curing. Barrier characteristics of film were enhanced through lowering permeability and rising length of penetration path for O_2 and water. Scattering of clay NPS among the coating matrix was revealed through SEM and TEM. PBZ/E-SBO/nanoclay (3 wt%) compound looks appropriate for the application of interface coating of mild steel that shows excellent outcomes as corrosion inhibition purpose.

Amany M. Fekry et al[58]; Reported a novel electrochemical sensor and that is too for developing nano-TiO_2/graphene oxide (TONP/GO).They have developed this sensor to screen the corrosion rate of nano-layered surgical biocompatible 316 L stainless steel implant among imitated body fluid (PBS) having 10^{-3} M ibandronate sodium (IBN) medicine that behaves as a corrosion resistance drug in pH 7.4, at

37^0C. Their new organic polymer/inorganic /ceramic nano-composite contains chitosan (CS), TiO_2 nano-particle (TONP), hydroxyapatite nano-particle (HANP), and multi-walled carbon nanotube (CNT). TO examine its structural property, corrosion behavior, and other characteristics they had used methods like potentiodynamic polarization, cyclic voltammetry (CV), electrochemical impedance spectroscopy (EIS), scanning electron microscope (SEM), and energy dispersive X-ray analysis (EDX). This synthesis material is biocompatible and very well reveals its capacity to enhance structural characteristics like robustness, corrosion resistance, and many more for the alloy. SEM images of (A) CS/TONP and (B) CS/TONP/HANP/CNT nano-layered surgical 316L stainless steel (SS) implant is shown in figure 1, respectively. CS/TONP layer diffused like convergent small balls among the chitosan matrix for TONP (figure.1A). Their thick coating provides excellent durability to surgical 316L SS alloy. Dual magnification in CS/TONP/HANP/CNT nano-film reveals through figure-1B. In an initial intensification of about 40000X exhibits homogenous CNT scattering along with TONP, HANP among the chitosan matrix. Through this homogeneity, the composite structure will have efficient improvement among its mechanical and tribological characteristics. They show the next intensification of 10000X that in picture exhibits that film appears to be a broccoli shape providing a thick film of good durability as compared to CS/TONP film. Enhancement of scattering of TONP, HA, CNT in the film is because of chitosan. ED analysis was shown in Figure 1C for the nanocomposite coating that proves the occurrence of every component over alloy interface in an effective amount that is a carbon (47.4%), oxygen (26.02%), nitrogen (2.53%), calcium (7.88%), phosphorous (0.96%), and titanium (0.86%). The highest percentage of CNT supports fine rupture strength and corrosion resistance of HANP, TONP supports HANP toughness towards interface also CS facilitates rising absorption characteristics of the layer through enhancement of dispersion over the alloy.

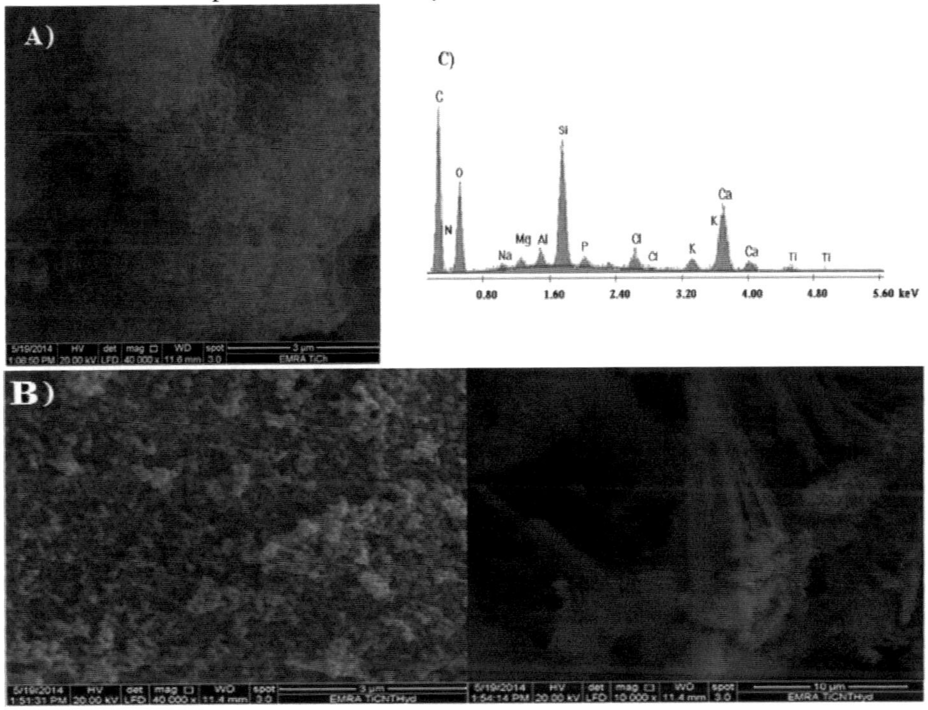

Figure 1. SEM pictures for (A).CS/TONP, (B). CS/TONP/HANP/CNT and (C). EDX analysis for CS/TONP/HANP/CNT nano- layered surgical 316L SS implant[58].

Moreover, the polarization graph (figure-2) to show non-coated surgical 316L implant (SS), SS/CS/TONP, and CS/CS/TONP/HANP/CNT the following immersion in PBS for 7 days with occurrence and non-occurrence of 10^{-3}M IBN drug at 37^0C. Physical quantities of polarization that are corrosion current density (i_{corr}) and corrosion potential (E_{corr}) can be derived from the Tafel extrapolation polarization curve. Corrosion current density comes out to be 20.1 nA cm^{-2} for pristine alloy, 5.01 nA cm^{-2} for CS/TONP nanolayer and falls to 1.7 nA cm^{-2} for CS/TONP/HANP/CNT nanolayer, this shows less corrosion rate is because of corrosion current value was directly dependent on corrosion rate. Afterward, they calculated the corrosion resistance efficiency of CS/TONP/HANP/CNT nano-film with 10^{-3} M IBN drug absorption to be 99.0% and if the drug is not there then it comes out to be 91.6% in comparison with 75.1% for CS/TONP/ nano-film.

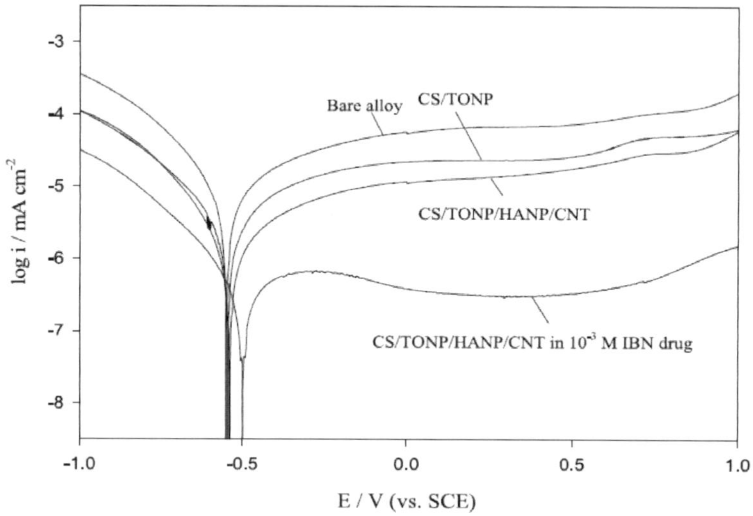

Figure 2. Polarization curve for pristine, CS/TONP, and CS/TONP/HANP/CNT nanolayered, surgical 316L SS implant following immersion for 7 days in PBS for occurrence and non-occurrence of 10^{-3} M IBN drug at 37^{0}C .[58]

Also, an electrochemical based sensor having TONP/GO is employed to get peak current with a variation of time for nano-film layered alloy, following the immersion in PBS having 10^{-3} M of IBN drug. It verifies that current falls with time make sure the rise in drug absorption over the examined interface that leads to a lesser corrosion rate.

Ruifeng Zhang et al[59]; Reported the development of Polyaniline (PANI) nano-fiber through a chemical oxidative polymerization technique. Styrene block copolymer polystyrene-b-(ethylene-co-butene)-b-styrene (SEBS) and SEBS/PANI fibers were also reported through the electrospinning development technique. Fiber structural deformation was examined through tetrahydrofuran (THF)/toluene binary solvent. PANI got charge and shifts towards the interface of Taylor cone as a high voltage electrostatic field is applied. The unique micro-nano fiber structure of SEBS/PANI is developed as THF evaporates fast, then PANI nano-fiber was piled up and implanted over the interface of the SEBS structure. They had examined the morphology of composite through SEM, TEM, UV-vis spectra, FTIR, and Raman spectra. Synthesized PANI reveals regular interface and small nanocomposite deformation having a diameter of about 40-50 nm with length 700±12 nm. PANI gives enhanced capability towards chemical sensing, liquid dispersion, and conductivity as it is placed in heat or flash. That gives radiation, which is held responsible for chemical cross-linking among PANI composite .developed electrospun coating has efficient strain's revival ability, also minute quantity of PANi enhances the heat stability of composite coating in comparison with SEBS coating. Electrospun SEBS/PANI coating had excellent corrosion inhibition for stainless steel. On bearing tensile strength more than 300% it still exhibits corrosion inhibition. Generally employed polarized Tafel graph was employed here to examine corrosion outcomes. Through the connection of relative anodic and cathodic polarization corrosion current (I_{corr}) and corrosion potential (E_{corr}) were calculated. Lesser I_{corr} value means less corrosion and large E_{corr} means the substrate is tough towards corrosion through the destructive medium. Bare 304SS

stainless steel had 1.943x10^{-6} A/cm^2 and 0.490 V I_{corr} and E_{corr} value, respectively with the rate of corrosion V_{corr} comes out to be 2.7x10^{-2} mm/year. V_{corr} Value for spin-coated SEBS was 1.5x10^{-2}, for electrospun SEBS it was 2.1x10^{-3} mm/year and for electrospun SEBS/PANI it was 4.0x10^{-4} mm/year. Electrospun SEBS/PANI had the largest E_{corr} value and the lowest value of I_{corr}, V_{corr} that reveals its eminent behavior as a corrosion inhibitor. They also showed that interface morphology and hydrophobic behavior varies corrosion inhibition. The water contact angle of SEBS was 136^0 and SEBS/PANI had a water contact angle 134^0. This was much larger than the SEBS contact angle that is 100^0. Irregular micro-composite efficiently inhibits the rise in the angle of contact of coating, also hydrophobic interface inhibits electrolyte interruption that shows excellent corrosion inhibition.
I Pugazhenthi et al[60]; Reported corrosion resistance characteristics of poly(4-methyl-5-vinylthiazole) PVTZ membrane layer over mild steel. Moreover, the membrane property of corrosion inhibition was improved through the addition of Titania nanoparticles (TiO_2 NPs). Characterization of nano-composite was done by XRD, TEM, SEM, and UV spectroscopy. SEM micrographs shown in figure-3 reveals PVTZ and PVTZ/TiO_2 the composite had irregular and non-uniform deformation that proved the impact of TiO_2 NPs over polymer matrix. In addition to this, TiO_2 NPs had fine dispersion and well-matched with the matrix. They had analyzed finely dispersed TiO_2 NPs having approximate size 25 nm.

Figure 3. FESEM of (a) PVTZ, (b) PVTZ/TiO_2 (0.5), (c) PVTZ/TiO_2 (1.0), (d) TiO_2 NPs (2.0)[60]
The corrosion inhibition ability of PVTZ and its TiO_2 composite was observed through potentiodynamic polarization and electrochemical impedance spectroscopy in 3.5 (w/v) NaCl, which reveals the excellent performance of membrane having corrosion inhibition with UV restriction characteristics in insistent media. Nano-composite shows better corrosion inhibition in comparison with a polymer film. Finely scattered NPs among PVTZ matrix decrease tiny holes and give tortoise structures that restrict or interrupt electrolytes to reach the steel interface. So, PVTZ organic membrane provides efficient barrier characteristics and excellent corrosion inhibition that could develop a new pathway for the paint industry.
Harish Kumar et al[61]; Reported the sol-gel method to develop Cu, Ni, and Zn metal oxides (CNZMO) nano-particles mixed with polymer Polyaniline (PANI) which was prepared via controlled temperature

chemical preparation technique. Moreover to develop the CNZMO-PANI nano-composite in-situ method was employed. For characterization of CNZMO nano-particle, polymer and metal oxide polymer nano-composite UV-visible, FTIR, FESEM, XRD, and TGA/DTA methods were employed. IR results confirm the homogenous dispersion of $CuCl_2.2H_2O$, $NiCl_2.6H_2O$, and $ZnCl_2$ among matrix element that is PANI. Thermogravimetric (TGA) analysis shows that the Polymer nano-composite was thermally steady even at 670^0C. XRD analysis reveals the crystalline structure of synthesized polymer nano-composite. FESEM pictures shown in figure-4 affirm the homogenous dispersion of metal oxide nano-particle among polymer matrix PANI. The constituent range of metal nano-particles was around 50-80 nm. CNZMO captured inside the cavity of the benzene ring expands a little in PANI converting it in sponge type substance. Further, Corrosion inhibition characteristics of CNZMO-PANI nanocomposite was examined with 0.1N hydrochloric acid over mild steel and comes out to be excellent anti-corrosive material even at 80 ppm concentration.

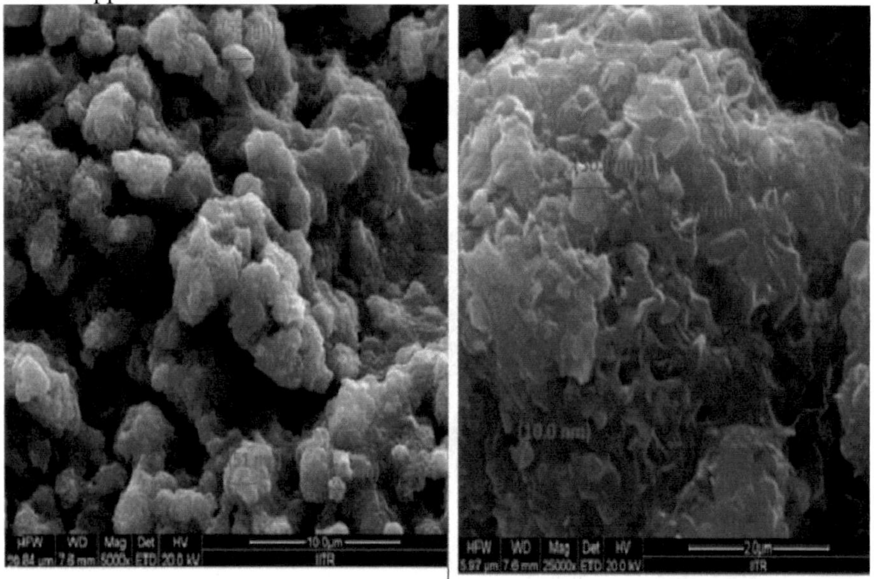

FESEM images of metal oxide nanoparticles	FESEM images of CNZMO-PANI nanocomposite.

Figure 4. FESEM pictures of metal oxide nanoparticles (left) and CNZMO-PANI nano-composite (right)[61].

Rajkumar Rajendran et al[62]; Reported chemical oxidative polymerization to synthesize poly(o-anisidine) (POA) nano-particle with silicon dioxide (SiO_2) nano-composite as POA/SiO_2 having various constituent substance of SiO_2 (5%, 10%, & 15% w/w). FT-IR study reveals that silicon oxide nano-particle (SNPs) and POA chains have specific interactions among each other. SEM (scanning electron microscopy) was employed for the study of interface deformation of polymer nano-composite that reveals the homogenous scattering of SNPs in the POA matrix and enhances the compact composition of the POA. XRD criteria of the POA, SiO_2 nanoparticle and POA/SiO_2 nanocomposite was shown in Figure-5. The approximate size of constituents was calculated through the Debye-Scherer equation and that comes out to be 5 nm. Figure-5b illustrates the characteristic peak of SNPs that reveals sturdy, broad peaks at center 2θ of 22.45^0 among different reports that exhibit the crystalline structure of SiO_2. Diffracted peaks of POA can be seen at 17^0 and 24^0 (Figure-5a) that shows the amorphous structure of POA. Broad peaks occur from amorphous polymer at 2θ near 24^0 falls and peak signifying protonated polymer comes into play that shows doping of POA among SiO_2 raises crystalline nature. One can notice diffraction peaks of POA and SiO_2 from Figure-5c for POA/SiO_2 composite that illustrates the merger of SNPs over the interface of the POA. Moreover, the intensity of the diffraction of the composite was not that strong as pristine SNPs. The approximate size of the constituents of synthesized POA nanoparticle and POA/SiO_2 nanocomposite comes out to be 38 nm and 20 nm respectively. In addition to this, the prepared composite material was initiated in acrylic resin than that polymer/SiO_2/acrylic hybrid composite was employed over the interface of the steel substrate.

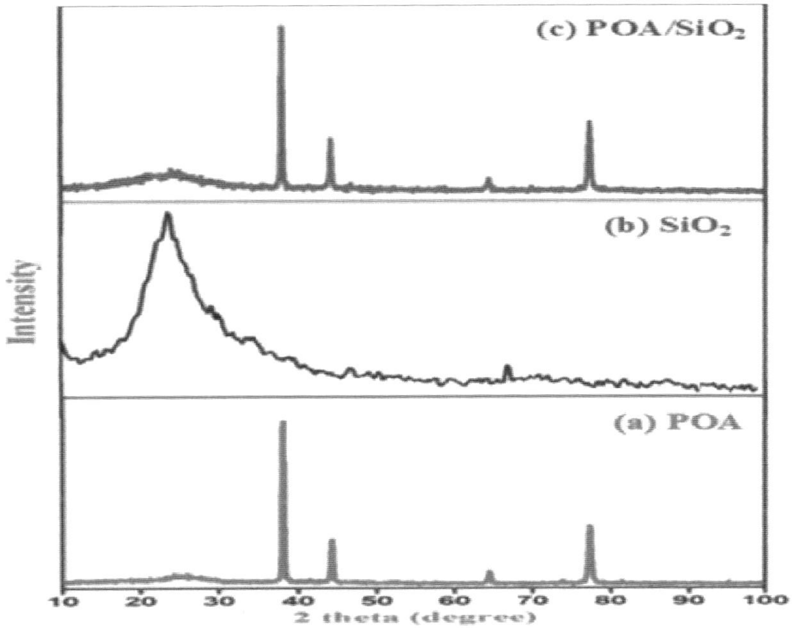

Figure 5. XRD criteria of (a) POA nano particle, (b) SNPs, and (c) POA/SiO_2 (15%) nano composites[62].

Polarization curve for acrylic layered and acrylic resin having various wt% content of POA nano-composite layered over mild steel samples (5%, 10% & 15%) with 3.5% NaCl mixture electrolytes. Tafel curve (fig-6(b)) was used to calculate various electrochemical quantities such as corrosion potential (E_{corr}), corrosion current density (I_{corr}), corrosion rate (C.R), and prevention efficiency (P.E %). Steel sample covered with 5-15% SiO_2 employed polymer composite in acrylic has an efficient rise in the calculated value of E_{corr}.

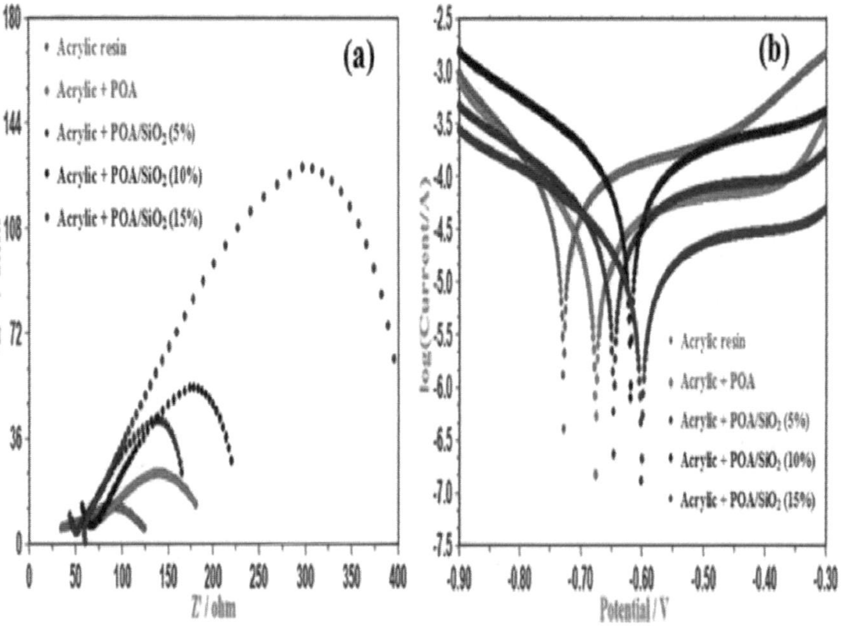

Figure 6. (a) Nyquist plot and (b) Tafel plot in 3.5% NaCl mixture after mixing POA and its nano-composites with the acrylic film over a mild steel sample.[62]

To analyze corrosion inhibition behavior of coated steel specimen P.E% was employed. From polarization values, it's evaluated that P.E of POA/acrylic, POA/SiO_2/acrylic (15%) composite film were 25.31% and 79.16% with 3.5% NaCl mixture respectively. So, it's clear that the incorporation of SNPs among POA enhances the corrosion resistivity of polymer nanocomposite along with acrylic film over mild steel specimens.

Guangyu Wang, et al[63]; Reported SiO_2/TiO_2 composites prepared with KH-570 enhanced nano-titanium dioxide and silica via in-situ polymerization process and achieve waterborne nano-hybrid agitated chained acrylic emulsion coating. The characterization process was done through FTIR (Fourier transform infrared spectroscopy), thermo-gravimetric analysis (TGA), and X-ray photoelectron spectroscopy (XPS).composition and deformation was examined through scanning and transmission electron microscope (SEM, TEM). The waterborne hybrid coating was characterized for corrosion resistance through potentiodynamic polarization and electrochemical impedance, salt spray test, wet adhesion test. The impedance of waterborne nano-hybrid hyper branch acrylic membrane comes out to be 8338Ω cm^2 in comparison with a pristine coating having an impedance 333Ω cm^2 membrane had much larger impedance. After extrapolating the polarization curve they got corrosion potential (E_{corr}) 0.85 µA/cm^2. It was noted that the corrosion current density (I_{corr}) decreases by tough compression of nano agitated chain composition. Salt spray test and wet adhesion tests were done at the end after coating composite material over less carbon steel substance. The time for the salt spray test goes beyond 1000h also wet adhesion reveals that the membrane achieves exceptional resistance in a moist medium.TGA at heating range 10^0C/min with nitrogen flux was done in examining the thermal decomposition of pristine acrylic emulsion and waterborne nanohybrid agitated chained acrylic emulsion. The first thermal degradation temperature (T_d) of the polymer phase refers to the temperature at which 5% of mass vanishes. T_d For pristine acrylic emulsion and nanohybrid composite comes out to be 297.3 and 335.8^{0}C respectively. Thermal adhesion comes out to be 38.5^{0}C larger in the nanohybrid composite than pristine acrylic emulsion. They characterized these values with the interaction of nano-titanium oxide and silicon oxide and acrylic emulsion. Their results revealed that waterborne nano-hybrid agitated chained acrylic emulsion was significantly useful for the rising temperature membrane. Basically, they had successfully

developed waterborne nano-hybrid hyper branch acrylic emulsion film which was eco-friendly, shows corrosion inhibition ability and had specific metal protection applications too
Niteen Jadhav et al[64]; Reported conducting polymer nano-composites polypyrrole (PPy)/Fe_2O_3 for corrosion inhibition over the cold spin substrate. Characterization techniques used for this composite were electrochemical impedance spectroscopy and open circuit potential evaluation. Salt fog experiment conditions were exposed to ASTM B117 for 40 days. Metal oxide and metal oxide/PPy film resistance graphs were drawn for the period of exposing specimens in salt spray (Figure-7). At the starting point, R_c (corrosion resistance) was large for metal oxide/PPy film in comparison with a metal oxide film. Resistance decreases with the time but manages to rise for the nano-composite film for a similar period of exposure. This shows that PPy/Fe_2O_3 composite film gives exceptional corrosion inhibition as compared to metal oxide coating Enhanced outcomes of coating is due to interface bonding of ions along PPy and passive film development of PPy

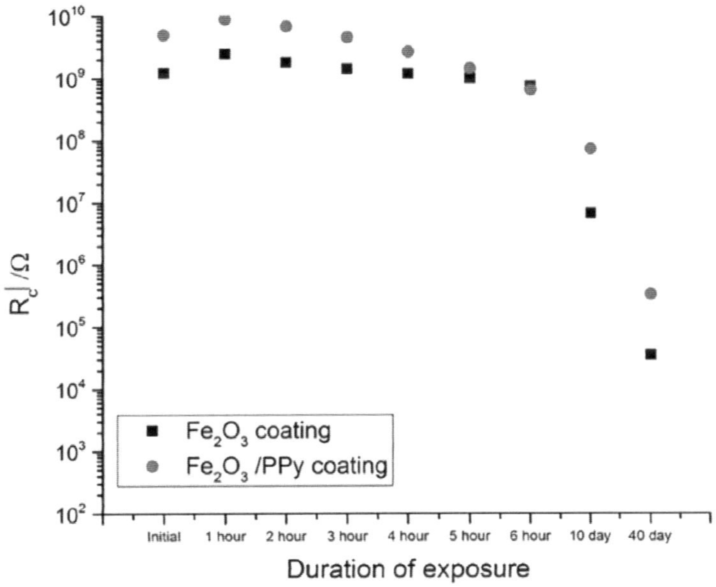

Figure 7. variation in resistance of coating (R_c) with the time of exposure[64].
Khulood A.Saleh et al[65]; reported the preparation of poly N terminal tetrahydrophthalamic acid on stainless steel 316 (SS) via electrochemical polymerization technique. Methods employed to characterize synthesized polymer composite was atomic force microscopy (AFM), Fourier transforms infrared spectroscopy (FTIR), and scanning electron microscopy (SEM). They have added nano ZnO and graphene to enhance the properties of polymer coatings. The anti-corrosion performance of pristine and coated SS was done by electrochemical polymerization with a 0.2 M hydrochloric acid (HCl) mixture. With the rise in temperature corrosion current density (I_{corr}) and corrosion potential (E_{corr}) go up. As they coat SS with polymer I_{corr} value falls with or without nano-material. Tafel plot was drawn and that shows E_{corr} Value of polymer coated SS with or without nano-material increases more in comparison with pristine SS, which exhibits an anodic protective characteristic of the polymer membrane. Moreover, protection efficiency (P.E%) was found to be large in the case of SS coated with polymer composite mixed with graphene as compared to a polymer composite mixed with ZnO. Further, they tested polymer film for anti-microbial activity against bacteria (S.aureus and E. coli), nano-material modified polymer composite exhibits robust activity for these bacteria.
K. Vinothkumar et al[66]; reported composite coating consist of graphene oxide (GO) and electro-polymerized p-3-amino-5-mecapto-1,2,4-triazole (p-AMTa).composition and morphology was

characterized with FTIR (Fourier transform infrared spectroscopy), Raman spectroscopy, XPS (X-ray photoelectron spectroscopic analysis), and energy-dispersive X-ray spectroscopy (EDX). They have used electro-polymerization of AMTa adding GO in it to form composite material over the copper substrate. The formation of S-S and metal sulfur connections was revealed via FTIR. N-O broadening frequency confirmed the bonding among electrophilic carbon atom of graphene oxide and the nitrogen atom of Azoles. The addition of graphene oxide among p-AMTa was well confirmed with D and G bands in Raman analysis. SEM study reveals for (a) metallic copper, (b) p-AMTa/GO composite layered copper, (c) pristine Cu, (d) p-AMTa/GO composite layered over Cu immersed in NaCl (3.5%) for three days (Figure-8).

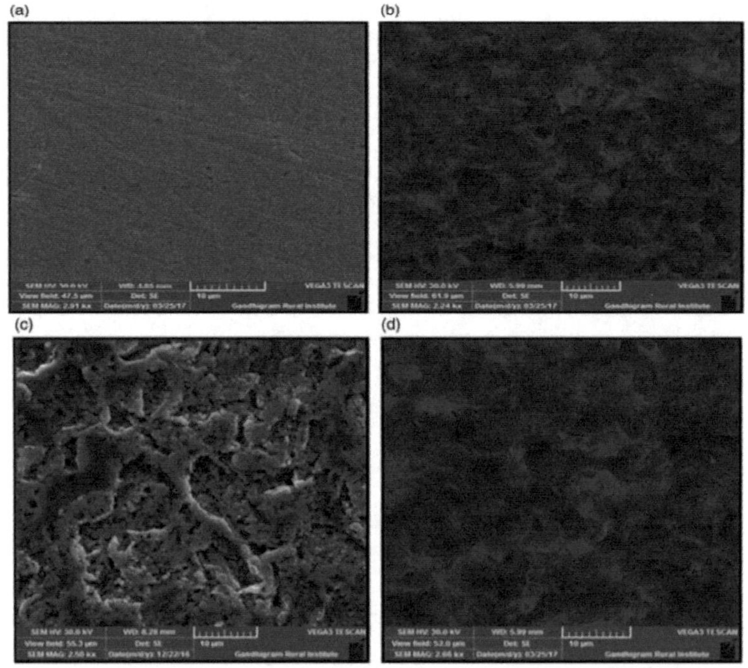

Figure 8. Interface deformation of (a) pristine Cu, (b) p-AMTa/GO composite layer over Cu, (c) pristine Cu, and (d) P-AMTa/GO composite coated over Cu with three days immersion in NaCl solution (3.5%) [66].

SEM analysis confirming uniform cloudy film of polymer composite layered on the copper substrate. Pristine copper corroded badly when immersed for 3 days in 3.5% NaCl solution (Figure-8c). But on immersing polymer composite that is p-AMTa/GO material in NaCl mixture no effect of corrosion was observed over the interface (Figure-8(d)) this is due to the formation of an anti-corrosion layer with polymer composite. EIS (electrochemical impedance spectroscopy) and PDS (potentiodynamic polarization studies) were used to test the corrosion resistance of composite material under the action of NaCl mixture and confirmed that prevention of anodic mixture of Cu was efficiently done by p-AMTa/GO composite. Moreover, bode graphs for Cu, p-AMTa, and p-AMTa/GO composite reveals dual maxima pointing dual constants for Cu, at 10^2 Hz, and 10^2 Hz different frequency. The corrosion resistance efficiency of polymer graphene oxide composite material was coming out to be 78.48%.

Helma Sadat Bahari et al[67]; Reported chitosan nano-composite coating over Cu to examine its corrosion resistance ability. The coating exhibits excellent thermal stability with the incorporation of silica nanoparticle (NP) and 2-mercaptobenzothiazole (MBT) that decreases the swelling proportion of chitosan. Chitosan was self interlinked with adding one element to another with interconnects such as tripolyphosphate (TPP) and for covalent connection glutaraldehyde (GA). Barrier protection of chitosan coating was enhanced via MBT and silica incorporation among matrix elements, a rise in corrosion inhibition was analyzed through EIS and maximum corrosion resistance efficiency comes out to be 85%. The interconnected Chitosan coating was better than non-connected coating elements for corrosion inhibition. The corrosion mechanism shows the mass transition and dispersion phenomenon. Comers'-kronig transformation confirmed the experimental calculation of impedance, also polarization graph validates impedance. Corrosion inhibition from a waterborne polymer film mechanism was connected to a micro-porous composition and solvent absorption among film. Swelling of polymer, caused by solvent molecules dispersed in polymer from the mixture was the significant property of polymer film to calculate inhibition efficiency over metal corrosion. The evaluated swelling ratio or solvent resistance of chitosan-based film synthesized in this report was given in table-1 giving the reference of non-interconnected and interconnected compositions. Which was well analyzed in this study.

Table 1. Swelling ratios of interconnected and non-interconnected chitosan coating over Cu was calculated via the quantity of water uptake. The approximate value of ratios of three replicates for every film is given below[67]

Non-interconnected	Swelling ratio	Interconnected	Swelling ratio
CS	2.37	CS-GA	0.77
CS-MBT	1.23	CS-MBT-GA	0.80
CS-SiO_2 2wt%	0.62	CS-SiO_2 2wt%	0.95
CS-SiO_2 5wt%	0.80	CS-SiO_2 5wt%	0.80
CS-SiO_2 10wt%	0.88	CS-SiO_2 10wt%	0.83
CS-MBT-SiO_2 2wt%	0.67	CS-MBT-SiO_2 2wt%	0.46
CS-MBT-SiO_2 5wt%	0.28	CS-MBT-SiO_2 5wt%	0.68
CS-MBT-SiO_2 10wt%	0.54	CS-MBT-SiO_2 10wt%	0.56

Incorporating a large quantity of silica NPs resist the efficient interconnection of chitosan because of electrostatic interaction among amino chains. So silica with interconnected chitosan reveals very low coating inhibition. For effective identification of the consequence of silica NPs, membrane inhibition with varying silica NPs value for various chitosan interconnected and non-interconnected films reveal in the image (Figure-9).

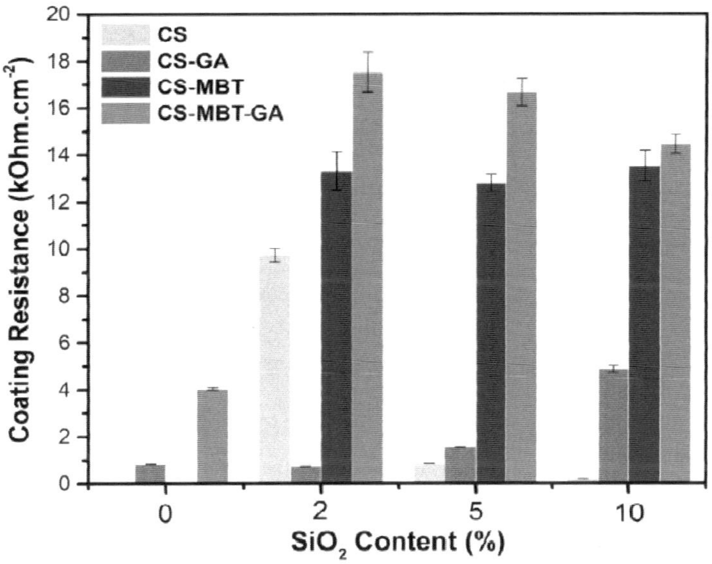

Figure 9. Coating inhibition evaluated for various chitosan compositions with a variation of SiO_2 [67]

Samarah Vargas Harb et al[68]; Reported effective results for the Ti6Al4V metal alloy problem of bad durability for corrosion inhibition and osseointegration. They have developed poly (methyl methacrylate)-silicon dioxide (PMMA-silica) film for Ti6Al4V alloy as a corrosion inhibitor as well as a bioactive membrane. They employed the sol-gel method incorporated with radical polymerization for PMMA interconnected with covalent bond with silica phase via 3-(trimethoxysilyl) propyl methacrylate (MPTS). They obtain a thickness of 5μm for coating, moreover a uniform, apparent, non-porous and cracks free with excellent linkage for the metallic surface that is 11.6 MPa. The coating was characterized for anti-corrosion application through electrochemical impedance spectroscopy (EIS) and results were eminent showing corrosion inhibition 26 GΩ cm^2 and sustaining stability among simulated body fluids (SBF) mixture for more than 100 days. The excellent corrosion resistance of PMMA-silica film was clear through phase angle plot showing angle near to 90^0 nearly for whole frequency variation. This property was eminent for a quasi-ideal capacitor that can obstruct the attack of electrolytes in a coating that inhibits corrosion for metal surfaces. Their outcomes tell that PMMA film provides an only modest enhancement in corrosion inhibition but silica incorporation gives effective enhancement for corrosion protection and makes it durable for a longer time. Further for 21 days of immersion in SBF, PMMA-silica film reveals apatite layer attributes bioactivity of the composite. Bioactivity characterization reveals improvement in osseointegration which was due to rising interface free energy and absorption of protein. So basically, the synthesized PMMA-silica coating for Ti6Al4V alloy was eminent for both corrosion resistance and bioactivity.

Gallegos-Melgar et al[69]; Reported green synthesis of composite film through heating sulfuric anodizing for preparing Al_2O_3 transform coating containing organic chitosan-sodium alginate incorporated coating via electro-deposition method. Following the inorganic phase (Al_2O_3) and organic coating's micro composition was characterized by scanning electron microscopy (SEM). In Anodic aluminum oxide (AAO) coating to identify pores, cavities, defects, and deformation of composition SEM was used. We know that corrosion inhibition of anodized aluminum relies on coating depth and the last outcome of the preserved interface. So anticipation is an important focus of polymer film was to preserve interface, As a result, the corrosion inhibition increases. Potentiodynamic polarization curve was employed to evaluate corrosion inhibition of bare and coated AA6063 alloy under imitated sea atmosphere and then the comparison of the result was done with specimen immersed in similar electrolyte for 30 days. Chitosan coated specimen reveals high polarization efficiency; the chitosan-alginate film specimen shows passivation exhibiting distinction from chitosan film, this was due to intermixing or merging two polymers that form polyion complex (PIC) due to the presence of ionic cross-linking. Their important corrosion mechanism anticipated protonated carboxylates and amino chains were coated on cathodic and anodic positions, respectively while depositing and potentiodynamic polarization curve due to aluminum ions improves cathodic reaction. So, the outcomes of this research exhibit probability of employing prepared corrosion resistance film for inhibiting aluminum alloy from chloride corrosive ions.

Yanning Chen et al[70]; Reported the development of a new structural composition like a sandwich for corrosion inhibition application. They have fabricated eco-friendly composition through depositions 8-hydroxyquinoline (8-HQ) over graphene oxide interface (GO).8-HQ, GO/8-HQ was improved by the incorporation of polydopamine (PDA) to avoid corrosion species to escape into the interface. Sandwich composition merges in the matrix element provide long-lasting durability of inhibitor. 8-HQ adding in GO interface via ring-opening reaction then, PDA network incorporated through dopamine self polymerization enhance the complexity of GO/8-HQ and form a sandwich composition of GO/8-HQ/PDA. The selection of PDA was done for improvement as it can be layered over the interface of the substrate via auto polymerization. GO nanosheets were exploited like distinctive flexible material and polymer groups of PDA keep long-lasting reliability of composition. To validate the composition and deformation of the structure of composite material Fourier transform infrared spectroscopy (FTIR), X-ray photoelectron spectroscopy (XPS), and scanning electron microscopy (SEM) were employed. Further to test the corrosion resistance of GO/8-HQ/PDA-EP structure composition was held immersed in NaCl mixture for 140 days. The polarization graph and EIS are represented in figure-10 (A, B) respectively. Corrosion potential (E_{corr}) rise to -0.580 V and corrosion current (I_{corr}) decreases to 9.207x 10^{-8} A/cm^2 recorded on the 80th day of immersion (figure-10A). After the 80th day, it starts falling steadily in between 20-80 days it was rising steadily (figure-10B). Discharging 8-HQ from GO/8-

HQ/PDA encourages passivation behavior over the interface of AZ31b alloy and gives rise to corrosion efficiency among deficient points. So, Diagrammatical analysis of the specimen while corrosion resistance experiment reveals that the alloy interface layered with GO/8-HQ/PDA sandwich structure had a less quantity of interface corrosion and sustain as it is for about 140 days while leakage of corrosive ions. Adding well scattered GO/8-HQ/PDA among epoxy resin eminently enhances the corrosion resistance of AZ31b alloy that is demonstrated with an electrochemical experiment in 3.5 wt% NaCl mixture.

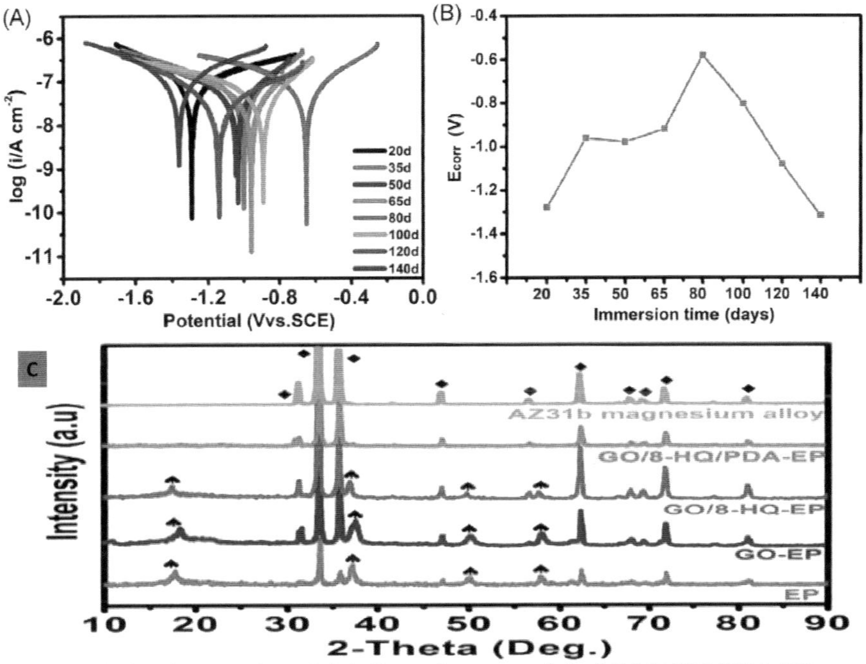

Figure 10. (A) Polarization graph and (B) Corrosion potential of GO/8-HQ/PDA-EP structure while 140 days of immersion, (C) XRD peaks of corrosion species below coating following immersion [70]

The pristine epoxy film shows less corrosion resistance as compared to the sandwich composition GO/8-HQ/PDA/EP structure. Sandwich composition shields the motion and compositional reliability of inhibitor 8-HQ.8-HQ hack ion-exchange among metal alloy and electrolyte mixture that obstruct electrochemical metal corrosion. The structure of the film after immersion of 140 days was analyzed by XRD (fig-10C). An important diffraction peak relates to the metal phase of the metal alloy substrate. Moreover, corrosion substance $Mg(OH)_2$ arises for GO/8-HQ/PDA-EP shows no peak, pointing that the path for the redox reaction going towards the metal corrosion interface was hacked. GO/8-HQ/PDA composition demonstrating enhanced corrosion inhibition as compare to GO/8-HQ and single GO composition, this is due to the inflexibility of GO towards corrosion electrolytes, and its structured composition shows the large surface area ability to deposit an ample quantity of 8-HQ. It's well known that 8-HQ can show the complexity for metal ions to hinder attacking corrosion species. If corrosion ions attack coating, discharge Mg^{2+} ions with 8-HQ anions are likely to organize and provide a complex structure for corrosion imperfection and that efficiently obstructing corrosion species. Moreover, GO/8-HQ/PDA-EP interface was integral in comparison with GO/8-HQ-EP having no corrosion species. Polymer PDA enhances hydrophobicity of structure and shows specific physical hindrance, that efficiently avoiding straight contact of NaCl mixture. Further, PDA provides a thick EP matrix element having minimum cracks and pores among coating material that resist corrosion species to flow into the surface. Hence, GO/8-HQ/PDA-EP sandwich structure having hydrophobic behavior shows better corrosion inhibition characteristics. Corrosion mechanism was explained and this analysis offer

possibility to restrict corrosion resistors over metal interfaces in an effective manner and can be employed for commercial purpose in the upcoming time
Min Min Aung et al[71]; Reported novel composite contains jatropha oil epoxy acrylate (AEJO) resin that acquires the ability to change coatings based on non-environment friendly methodologies. Epoxy acrylate mixed with nano zinc oxide and forms a hybrid nana composite for corrosion inhibition of mild steel surface. Various quantities of zinc oxide reinforcement (1, 3, 5, 7, and 9 wt %) were dispersed in the AEJO matrix. The scattering of ZnO filler among the epoxy matrix was analyzed with FT-IR spectroscopy. X-ray diffraction (XRD) was employed to observe diffraction characteristics and FESEM evaluate desired outcomes of the nano element with the polymer. Interface deformation of AEJO/ZnO composite film presented in Figure-11. A large surface area of inorganic constituents improves physical characteristics efficiently. In pristine coating porous deformation showing circular pores (Figure-11a). Inorganic reinforcement was scattered uniformly among matrix that improves the physical characteristics of the hybrid film. With 1 and 3 wt% AEJO/ZnO film range and quantity of pores decreases (Figure-11b, c). Negligible pores are revealed over the interface of the composite film with 5 wt% (Figure-11d). Excellent interaction among filler and organic matrix was demonstrated as ZnO was homogeneously scattered in 5 with% AEJO/ZnO hybrid film. With a larger quantity of ZnO, filler agglomeration occurs (Figure-11e) which decreases the homogeneous scattering of filler that attributes corrosion efficiency of the coating. The wetting phenomenon of ZnO film and contact angle (CA) was evaluated with contact angle meter (Figure-11 a-f). While analyzing it was determined that CA for pure specimen was 85.2 ± 2.01^0 and values enhances about 95 to 104.4 ± 2.01^0 as ZnO was incorporated. This is due to the hydrophobic behavior of ZnO maximize CA of the coating when dispersed in AEJO. CA showed a value larger than 90^0 which means water will not leak into the surface of the coating, therefore contributing towards better inhibition characteristic.

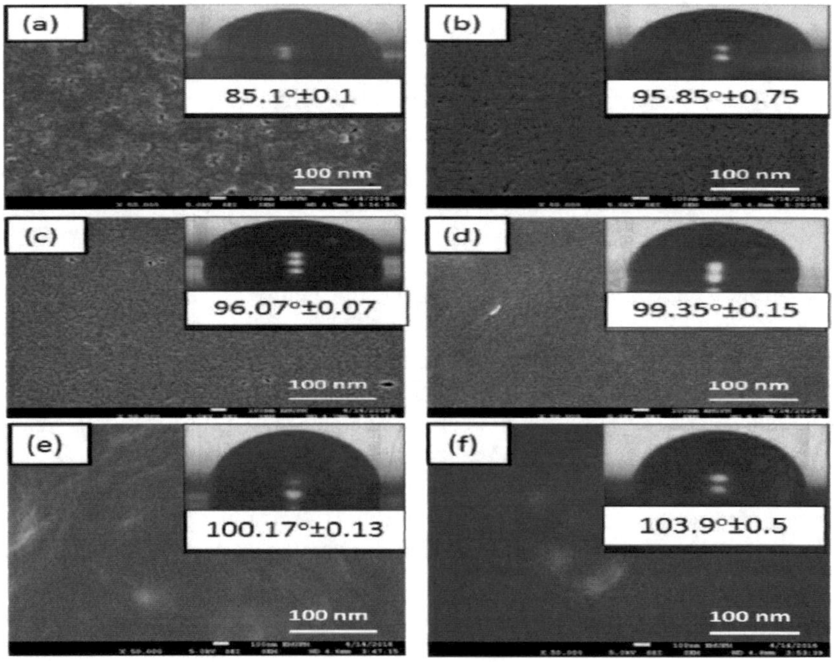

Figure 11. FE-SEM images of the (a) pristine coating, (b) AEJO-ZnO 1 wt %, (c) AEJO-ZnO 3 wt %, (d) AEJO-ZnO 5 wt %, (e) AEJO-ZnO 7 wt % and (f) AEJO-ZnO 9 wt % [71].

A corrosion resistance test was done for hybrid coating with electrochemical impedance spectroscopy (EIS), Tafel plots, and salt spray test. Hybrid coating immersed in 3.5 wt% NaCl mixture revealed enhanced performance of barrier protection property of coating. 1-3 wt% of ZnO film shows a larger corrosion field as compare to 5, 7, and 9 wt% of ZnO. They reported that coating inhibition enhances as

ZnO quantity and inhibition impact of ZnO in incorporated in a good manner with AEJO. In comparison with 0, 1, and 3 wt% of ZnO 5, 7, and 9 wt% shows enhanced corrosion efficiency. This was due to nanoparticle reinforcement that might reduce the penetration of electrolytes of organic coating hence reduces the rate of corrosion. This report gives eco-friendly composition and inhibition for surface and surface engineering.

4. **Major concerns**: There are some concerns related to development and applicability of polymer nano-composites as corrosion inhibitors faced by various researchers[72] are given below-

(a). Homogeneous scattering: In various reports, scattering components employed as reinforcing polymer matrix was not that successful to provide the required reinforcement that is because of agglomeration of the matrix. Homogenous scattering of matrix elements is promising for improved coating characteristics also the techniques for those are essential to investigate more particularly in case of high concentration matrix elements.

(b). New coating substance: A coating substance having immense mechanical durability is still needed to investigate. A substance with effective corrosion inhibition with less mechanical durability and vice versa. Synthesizing hybrid nana materials for coating showing potential characteristics is greatly essential for recent corrosion problem with no loss of interface mechanical behavior of the membrane.

(c). In-situ method: Not many reports are there for the in-situ method of corrosion, this method or analysis can provide great knowledge for corrosion progression[73], [74]. To get depth information about corrosion and its chemical properties electrochemical corrosion analysis needs to collaborate with the present interface investigation process[75].

(d). Modern techniques: Association among the micro composition of coating substance and its characteristics can be analyzed entirely with modern characterization methods like high-temperature nanoindentation, atom probe tomography, x-ray nano-diffraction, and many more[76].

(e). Commercial applicability: Currently various corrosion inhibitors giving improvised results are being employed and analyze by researchers. Moreover, many techniques are required for industrial establishment and feasibility. Novel methods are needed to explore during the starting phase in the internal system of the coating including corrosion-resistant substance.

(f). Ecological concern: Like all scientific researches it is necessary for coating substances also not to harm the ecological system directly or indirectly.

All in all, we illustrate challenges and concerns for polymer nanocomposites in two ways: protection and usability. Budding harmful properties of nanoparticles for the environment and humans is a topic of concern nowadays. The debate is over the artificial nanoparticles rising from its balanced state because of the maximum interface energy that makes it highly unsteady and approachable for transformation to acquire a comparative sustainable phase[77]. Hence, the material could transfer to the ecological system and interact with other materials or elements to acquire stability that can be hazardous for the environment and humans. Giving an example nanoparticles liberated in the environment can be absorbed through the roots of plants (figure-12) ejected itself in the plant's vascular system then towards stem and leaves. A.dev[78] and Jhao et al[79], [80]; explained such probable transformation of nanoparticles in plants revealing major concerns for the synthesis of polymer nanocomposites. Liang Xu et al; [81], illustrated sustainability, the interaction of nano-particles, transformation of nanoparticles in plants. Such novel advances and results comprise important pros and cons for the synthesis of polymer nanocomposite for corrosion inhibition.

Usability of polymer nano-composites for corrosion inhibition is not for atmosphere consists of gases such as CO_2 and H_2S over an immense rise in temperature[82]. Gases can simply disperse in polymer coatings over elevated temperatures that can increase the chances that corrosion species can harm the substrate surface. At such elevated temperature polymers composites chains starting to decay. Even their use as corrosion inhibitors was under temperature less than 120°C.

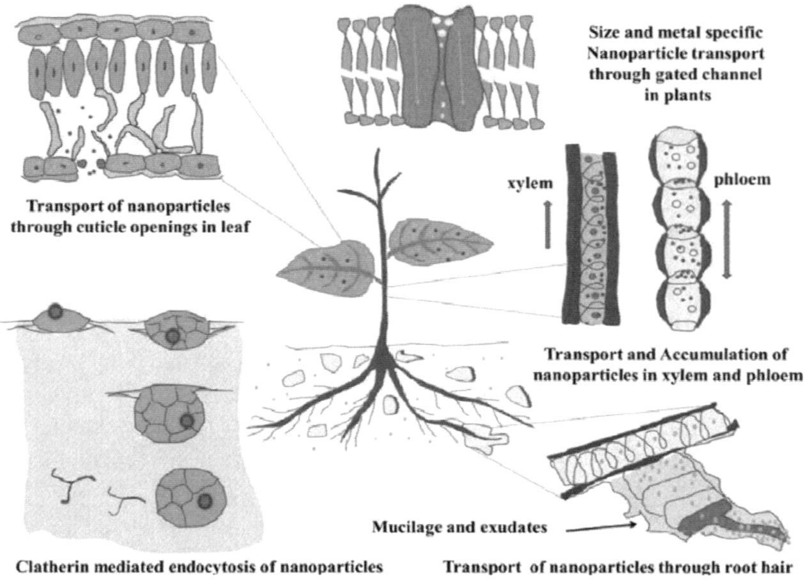

Figure 12. Uptake, growth, and transformation of nano-particles among plants[78].

Regardless of these issues, polymer nanocomposites provide reasonable techniques for corrosion resistance of metals. So it would be admirable for upcoming researchers and authors to spotlight synthesis methods that can extend the span of use of such materials for corrosion inhibition. From the review, it emerges that polymers those present in the environment naturally or we can say organic polymers are less employed. The range of such polymers is very much cost-friendly and environment friendly too[83] so they must be utilized completely. For the development of nanocomposite from polymers stability and interaction of nanoparticles with polymer matrix must be discussed and analyzed deeply.

5. Conclusion: Corrosion of metals is a huge concern for many commercial fields. To handle this problem, deep knowledge of corrosion progression over interface among corrosion electrolyte and metal interface is important. Well, designed film consist of nanocomposites is among popular and excellent techniques to protect metals from corrosion. In this review, we have discussed literature that shows that developed corrosion inhibitors can maintain the reliability of metal interfaces. Materials like Chitosan, PANI, and PMMA are widely used as polymer composite for corrosion inhibition. Moreover, some concerns like analyzing, in-situ corrosion techniques, and new methods of characterizations need to be discussed extensively. New composite substances and natural polymers need more spotlight. Hybrid nanocomposites for more than a single matrix as the coating substance can be more useful because the incorporation of more nanoparticles provides excellent and improved outcomes.

References

[1] Ge L, Li Q, Wang M, Ouyang J, Li X and Xing MMQ 2014 Nanosilver particles in medical applications: synthesis, performance, and toxicity *Int. J. Nanomedicine* **9** 2399–2407.
[2] Hebeish A, Hashem M, El-Hady MMA and Sharaf S 2013 Development of CMC hydrogels loaded with silver nanoparticles for medical applications *Carbohydr. Polym* **92** 1 407–413.
[3] Javidi M, Zarei M, Naghavi N, Mortazavi M and Nejat AH 2014 Zinc oxide nanoparticles as a sealer in endodontics and its sealing ability *Contemp. Clin. Dent* **5** 1 20–24.
[4] Souza VGL and Fernando A L 2016 Nanoparticles in food packaging: Biodegradability and potential migration to food-A review *Food Package. Shelf Life* **8** 63–70 .
[5] Sorrentino A, Gorrasi G and Vittoria V 2007 Potential perspectives of bio-nano composites for food packaging

applications *Trends Food Sci. Technol* **18** 2 84–95.
[6] Ali MA *et al 2014* Nanotechnology : A new frontier in Agriculture *Int. J. Adv. Life Sci* **1** 3 129–138.
[7] El-Bendary H and El-Helaly A 2013 First record nanotechnology in agricultural : Silica nanoparticles a potential new insecticide for pest control *Appl. Sci. Reports* **4** 3 241–246.
[8] Gensler R, Gröppel P, Muhrer V and Müller N 2002 Application of nanoparticles in polymers for electronics and electrical engineering *Part. Part. Syst. Charact* **19** 5 293–299.
[9] Bi S, Shi L and Zhang L 2008 Application of nanoparticles in domestic refrigerators *Appl. Therm. Eng* **28** 14 1834–1843.
[10] Anselmann R 2001 Nanoparticles and Nanolayers In Commercial Applications *J. Nanoparticle Res* **3** 4 329–336.
[11] Umoren S A and Solomon MM 2019 Protective polymeric films for industrial substrates: A critical review on past and recent applications with conducting polymers and polymer composites/nanocomposites *Prog. Mater. Sci* **104** 380–450.
[12] Barcikowski S, Hahn A, Kabashin AV and Chichkov B N 2007 Properties of nanoparticles generated during femtosecond laser machining in air and water *Appl. Phys. A* **87** 1 47–55.
[13] Luo X, Morrin A, Killard A J and Smyth M R 2006 Application of nanoparticles in electrochemical sensors and biosensors *Electroanalysis* **18** 4 319–326.
[14] Nagarajan R 2008 Nanoparticles: Building Blocks for Nanotechnology in *Nanoparticles: Synthesis, Stabilization, Passivation, and Functionalization* **996** American Chemical Society 1–2.
[15] Ammar S, Ramesh K, Vengadaesvaran B, Ramesh S and Arof AK 2016 A novel coating material that uses nano-sized SiO2 particles to intensify hydrophobicity and corrosion protection properties *Electrochim. Acta* **220** 417–426.
[16] Olad A and Nosrati R 2013 Preparation and corrosion resistance of nanostructured PVC/ZnO–polyaniline hybrid coating *Prog. Org. Coatings* **76** 1 113–118.
[17] cong Jiang C, ke Cao Y, Yong Xiao G, fu Zhu R and peng Lu Y 2017 A review on the application of inorganic nanoparticles in chemical surface coatings on metallic substrates *RSC Adv* **7** 13 7531–7539.
[18] Zheludkevich M L, Salvado I M and Ferreira M G S 2005 Sol-gel coatings for corrosion protection of metals *J. Mater. Chem* **15** 48 5099–5111.
[19] Asano H, Féron D, Gens R, Padovani C and Taniguchi N 2014 Long-term prediction of corrosion damage in nuclear waste systems *Corros. Eng. Sci. Technol* **49** 6 401.
[20] Schmitt GA *et al 2009* Global Needs for Knowledge Development in Materials Deterioration and Corrosion Control in cooperation with *Fed. High. Adm.*, vol. FHWA-RD-01, 1–44 .
[21] Sereda PJWeather Factors Affecting Corrosion of Metals in *Corrosion in Natural Environments*, W. H. Ailor, S. W. Dean, and F. H. Haynie, Eds. West Conshohocken, PA: ASTM International, 1974, pp. 7–22.
[22] Suzumura K and Nakamura SI 2004 Environmental factors affecting corrosion of galvanized steel wires *J. Mater. Civ. Eng* **16** 1 1–7.
[23] Fotovvati B, Namdari N and Dehghanghadikolaei A 2019 On coating techniques for surface protection: A review *J. Manuf. Mater. Process* **3** 1.
[24] Valipour NM, Birjandi FC and Sargolzaei J 2014 Super-non-wettable surfaces: A review *Colloids Surfaces A Physicochem. Eng. Asp* **448** 93–106.
[25] Jeon IY and Baek J B 2010 Nanocomposites derived from polymers and inorganic nanoparticles," *Materials (Basel).*, **3** 6 3654–3674.
[26] Quarta A, Di Corato R, Manna L, Ragusa A and Pellegrino T 2007 Fluorescent-Magnetic Hybrid Nanostructures: Preparation, Properties, and Applications in Biology *IEEE Trans. Nanobioscience* **6** 4 298–308.
[27] Zhang J *et al 2014* Sol-gel derived near-UV and visible antireflection coatings from hybridized hollow silica nanospheres *J. Sol-Gel Sci. Technol* **71** 2 267–275.
[28] Manasa S *et al 2017* Nanoclay-based self-healing, corrosion protection coatings on aluminum, A356.0, and AZ91 substrates *J. Coatings Technol. Res* **14** 5 1195–1208.
[29] 박병규 Byoung Kyeu TA-P, "자기치유 부식방지 코팅의 최근 동향 TT - Self-healing Coatings for Corrosion Protection: A Review of Recent Advances **47** 2014.
[30] Maity S and Chatterjee A 2013 Preparation and characterization of electro-conductive rotor yarn by in situ chemical polymerization of pyrrole *Fibers Polym* **14** 8 1407–1413.
[31] Suryanarayana C, Rao K C and Kumar D 2008 Preparation and characterization of microcapsules containing linseed oil and its use in self-healing coatings *Prog. Org. Coatings* **63** 1 72–78.
[32] Armelin E, Oliver R, Liesa F, Iribarren JI, Estrany F and Alemán C 2007 Marine paint fomulations: Conducting polymers as anticorrosive additives *Prog. Org. Coatings* **59** 1 46–52.
[33] Wang H, Xue Y, Ding J, Feng L, Wang X and Lin T 2011 Durable, Self-Healing Superhydrophobic and Superoleophobic Surfaces from Fluorinated-Decyl Polyhedral Oligomeric Silsesquioxane and Hydrolyzed Fluorinated Alkyl Silane *Angew. Chemie* **123** 48 11635–11638.
[34] Andreeva DV, Fix D, Möhwald H and Shchukin DG 2008 Buffering polyelectrolyte multilayers for active corrosion protection *J. Mater. Chem* **18** 15 1738–1740.
[35] Snihirova D, Liphardt L,Grundmeier G and Montemor F 2013 Electrochemical study of the corrosion inhibition ability of 'smart' coatings applied on AA2024 *J. Solid State Electrochem* **17** 8 2183–2192.
[36] Zheludkevich ML, Shchukin DG, Yasakau KA, Möhwald H and Ferreira MG S 2007 Anticorrosion Coatings with

Self-Healing Effect Based on Nanocontainers Impregnated with Corrosion Inhibitor *Chem. Mater* **19** 3 402–411.
[37] Michalik A and Rohwerder M 2005 Conducting Polymers for Corrosion Protection: A Critical View *Zeitschrift für Phys. Chemie* **219** 11 1547–1559.
[38] Fathima Sabirneeza A A, Geethanjali R and Subhashini S 2015 Polymeric Corrosion Inhibitors for Iron and Its Alloys: A Review *Chem. Eng. Commu* **202** 2 232–244.
[39] Winey K I and Vaia RA 2007 Polymer *Most* **32** 314–322.
[40] Vaia R A, Vasudevan S, Krawiec W, Scanlon L G and Giannelis EP 1995 New polymer electrolyte nanocomposites: Melt intercalation of poly(ethylene oxide) in mica-type silicates *Adv. Mater* **7** 2 154–156.
[41] Triantafillidis CS, LeBaron PC and Pinnavaia T J 2002 Thermoset Epoxy–Clay Nanocomposites: The Dual Role of α,ω-Diamines as Clay Surface Modifiers and Polymer Curing Agents *J. Solid State Chem* **167** 2 354–362.
[42] Fukushima Y, Okada A, Kawasumi M, Kurauchi T and Kamigaito O 1988 Swelling behavior of montmorillonite by poly-6-amide *Clay Miner* **23** 1 27–34.
[43] Kato C, Kuroda K and Misawa M 1979 Preparation of Montmorillonite-Nylon Complexes and Their Thermal Properties," *Clays Clay Miner* **27** 2 129–136.
[44] Kami M, Kampus K and Lumpur K 2013 Universiti Putra Malaysia Universiti Putra Malaysia 2–3.
[45] Pan YX, Yu ZZ, Ou YC and Hu GH 2000 New process of fabricating electrically conducting nylon 6/graphite nanocomposites via intercalation polymerization *J. Polym. Sci. Part B Polym. Phys* **38** 12 1626–1633.
[46] Khan MU, Reddy KR, Snguanwongchai T, Haque E and Gomes V G 2016 Polymer brush synthesis on surface-modified carbon nanotubes via in situ emulsion polymerization *Colloid Polym. Sci* **294** 10 1599–1610.
[47] Rahman I A and Padavettan V 2012 Synthesis of Silica Nanoparticles by Sol-Gel: Size-Dependent Properties, Surface Modification, and Applications in Silica-Polymer Nanocomposites—A Review *J. Nanomater* 132424.
[48] Cho JW and Paul D R 2001 Nylon 6 nanocomposites by melt compounding *Polymer (Guildf)* **42** 3 1083–1094.
[49] Shameli K, Bin Ahmad M, Yunus WZW, Ibrahim NA and Darroudi M 2010 Synthesis and characterization of silver/talc nanocomposites using the wet chemical reduction method *Int. J. Nanomedicine* **5** 743–751.
[50] Marin M L, McGilvray K L and Scaiano JC 2008 Photochemical Strategies for the Synthesis of Gold Nanoparticles from Au(III) and Au(I) Using Photoinduced Free Radical Generation *J. Am. Chem. Soc* **130** 49 16572–16584.
[51] M. Castillejo, P. M. Ossi, and L. V. Zhigilei, *Preface*, vol. 191. 2014.
[52] Novais S M V *et al 2018* Hydrothermal synthesis of CdWO4 for scintillator-polymer composite films development," *J. Lumin* **199** 225–231.
[53] Chen J, Wang J, Zhang X and Jin Y2008 Microwave-assisted green synthesis of silver nanoparticles by carboxymethyl cellulose sodium and silver nitrate *Mater. Chem. Phys* **108** 2 421–424.
[54] Salehoon E, Ahmadi S J, Razavi SM and Parvin N 2017 Thermal and corrosion resistance properties of unsaturated polyester/clay nanocomposites and the effect of electron beam irradiation *Polym. Bull* **74** 5 1629–1647.
[55] Quadri TW, Olasunkanmi L O, Fayemi OE, Solomon MM and Ebenso EE 2017 Zinc Oxide Nanocomposites of Selected Polymers: Synthesis, Characterization, and Corrosion Inhibition Studies on Mild Steel in HCl Solution," *ACS Omega* **2** 11 8421–8437.
[56] Solomon M M, Gerengi H, Kaya T and Umoren S A 2017 Enhanced corrosion inhibition effect of chitosan for St37 in 15% H2SO4 environment by silver nanoparticles *Int. J. Biol. Macromol* **104** 638–649.
[57] Mohamed M G, Kuo SW, Mahdy A, Ghayd I M and Aly KI 2020 Bisbenzylidene cyclopentanone and cyclohexanone-functionalized polybenzoxazine nanocomposites: Synthesis, characterization, and use for corrosion protection on mild steel," *Mater. Today Commun* **25** 101418.
[58] Fekry A M and Azab S M 2019 Nano-Structures & Nano-Objects The development of an innovative nano-coating on the surgical 316 L SS implant and studying the enhancement of corrosion resistance by electrochemical methods using Ibandronate drug *Nano-Structures & Nano-Objects*, no. XXXX, p. 100411.
[59] Zhang R *et al 2019* Corrosion Resistance of Stretchable Electrospun SEBS / PANI Micro-Nano Fiber Membrane," *Eur. Polym. J.*, p. 109394.
[60] Pugazhenthi I and Ghouse S M 2020 Corrosion protection performance of titania nanoparticles filled vinylthiazole) applied on mild steel in 3. 5 % sodium chloride solution.
[61] Kumar H, Boora A, Yadav A, Rajni and Rahul 2020 Polyaniline-metal oxide-nano-composite as a nano-electronics, optoelectronics, heat resistance, and anticorrosive material," *Results Chem* **2** 100046.
[62] Rajkumar R and Vedhi C 2020 Preparation, characterization and anticorrosion behavior of Poly(o-anisidine)-SiO2 nanocomposites on mild steel," *Mater. Today Proc.*, no. xxxx, 2020.
[63] Wang G, Wen S, Qian S, Wang J, Wang C and Chen Y 2020 Synthesis of novel nano hyperbranched polymer resin and its corrosion resistance in coatings *Prog. Org. Coatings* **140** 105496.
[64] Jadhav N, Kasisomayajula S and V. J. Gelling 2020 Polypyrrole/Metal Oxides-Based Composites/Nanocomposites for Corrosion Protection *Front. Mater* **7** 1–7.
[65] Saleh KA and Ali M I 2020 Electropolymerization for (N-terminal tetrahydrophthalamic acid) for anti-corrosion and biological activity applications *Iraqi J. Sci* **61** 1 1–12.
[66] Vinothkumar K and Sethuraman MG 2020 A robust method of enhancement of corrosion inhibitive ability of electrodeposited poly-3-amino-5-mercapto-1,2,4-triazole films over the copper surface using graphene oxide *J. Adhes. Sci. Technol* **34** 6 651–669.
[67] Sadat H *et al 2020* International Journal of Biological Macromolecules Chitosan nanocomposite coatings with

enhanced corrosion inhibition effects for copper *Int. J. Biol. Macromol* **162** 1566–1577.
[68] Harb SV *et al 2020* PMMA-silica nanocomposite coating: Effective corrosion protection and biocompatibility for a Ti6Al4V alloy *Mater. Sci. Eng. C* **110** 2019, 2020.
[69] I. Environment, "Potentiodynamic Polarization Performance of a Novel Composite Coating System of," 2020.
[70] Chen Y, Ren B, Gao S and Cao R 2020 The sandwich-like structures of polydopamine and 8-hydroxyquinoline coated graphene oxide for excellent corrosion resistance of epoxy coatings *J. Colloid Interface Sci* **565** 436–448 .
[71] Aung MM, Li WJ and Lim HN 2020 Improvement of Anticorrosion Coating Properties in Bio-Based Polymer Epoxy Acrylate Incorporated with Nano Zinc Oxide Particles *Ind. Eng. Chem. Res* **59** 5 1753–1763.
[72] Radhamani AV, Lau H, and Ramakrishna S 2020 Nanocomposite coatings on steel for enhancing the corrosion resistance: A review *J. Compos. Mater* **54** 5 681–701.
[73] Song Z and Xie Z-H 2018 A literature review of in situ transmission electron microscopy technique in corrosion studies *Micron* **112** 69–83.
[74] Fu X, Dong J, Han E and Ke W 2009 A new experimental method for in Situ corrosion monitoring under alternate wet-dry conditions *Sensors* **9** 12 10400–10410.
[75] Dwivedi D, Lepková K and Becker T 2017 Carbon steel corrosion: a review of key surface properties and characterization methods *RSC Adv* **7** 8 4580–4610.
[76] Tkadletz M, Schalk N, Daniel R, Keckes J, Czettl C and Mitterer C 2016 Advanced characterization methods for wear-resistant hard coatings: A review on recent progress *Surf. Coatings Technol* **285** 31–46.
[77] Xu L, Liang H-W, Yang Y and Yu S-H 2018 Stability and Reactivity: Positive and Negative Aspects for Nanoparticle Processing *Chem. Rev* **118** 7 3209–3250.
[78] Dev A, Srivastava A K and Karmakar S 2018 Nanomaterial toxicity for plants *Environ. Chem. Lett* **16** 1 85–100.
[79] Zhao L *et al 2013* Influence of CeO2 and ZnO Nanoparticles on Cucumber Physiological Markers and Bioaccumulation of Ce and Zn: A Life-Cycle Study *J. Agric. Food Chem* **61** 49 11945–11951.
[80] Zhao L *et al 2012* Effect of surface coating and organic matter on the uptake of CeO2 NPs by corn plants grown in soil: Insight into the uptake mechanism *J. Hazard. Mater* **225–226** 131–138 2012.
[81] Xu L *et al 2015* Understanding the stability and reactivity of ultrathin tellurium nanowires in solution: An emerging platform for chemical transformation and material design *Nano Res* **8 4** 1081–1097.
[82] Craig B D 1998 Selection guidelines for corrosion-resistant alloys in the oil and gas industry *Chem. Eng. World* **33**, 4 57–60.
[83] El-Haddad M N 2013 Chitosan as a green inhibitor for copper corrosion in acidic medium *Int. J. Biol. Macromol.*, **55**, 142–149.

Wireless body area network transmissions for iot-based healthcare network: a review

A Vinny Mary[1] and S Jerine[2]
[1]Department of Computer Science, Research Scholar, Noorul Islam Centre for Higher Education, Kumaracoil-629180
[2]Department of Software Engineering, Associate Professor, Noorul Islam Centre for Higher Education, Kumaracoil-629180

E.Mail:jamesvjj18@gmail.com

Abstract— This survey reviews how electronically aided health monitor ingand observing systems with wireless body area networks (WBANs) support patient's data analyzing and processing. These technologies in communications domain integrated into conventional medical system and assists as a capable methodology to boost the efficiency of health-care. Wireless body area networks contain dissimilar biosensors are fitted in a patient for analyzing the condition of health continuously. The collected parameters of signals are initially given at a gateway for the accretion of data and then furthered to out-of-the-way medical centers. Patient's smart devices which has internet connectivity act as the gateways in internet of things-based WBANs, each among them generally stands for single patient. A huge volume of medical data congestion will be occurred during the widespread use of IoT-based WBANs, which is leading to an imperative prerequisite with high utilization based efficiency system for radio resource management. Henceforth, this survey examines methodically the management of transmissions of wireless body area network (WBAN) for internet of things-based a network is sculpted as a multi-server priority queueing system with multi-class delay-constrained with. Based on the result of queueing data packet, a mechanism for the arrangement of data packet transmissions for medical purposes with delay constraints is deliberated.

Keywords— IoT-based healthcare, priority awareness, truthful mechanism, WBAN transmission scheduling

1. Introduction

With the improvements in technology which includes sensor technology, such as MEMS-micro-electro-mechanical systems, wireless based networking system, and low power consumption based sensors, there is a rapid improvement of wireless communication and low-strength multi-function sensors. The improvement on such technologies contributes to the development of such equipment portable, condensed and less-powered, offering a fundamental for the WBANs [1]. Wearable biosensor are one of the discipline of wireless body area networks or devices placed in a human body, which has emphasis on the dimension of such gadgets and less power wireless conversation among them. These nodes which have the sensor are capable of assembling physiological alerts of interest, which includes ECG data, blood hypertension and hypotension level, blood sugar level, temperature, voluntary and involuntary movement or activity indicators, and peripheral evidence of the human body, and also are able to process and broadcast of the acquired data.

Wireless body area network (WBAN) is one crucial module of electronic-health recording. Generally, a wireless body area network is arranged on a patient, and customarily contains of a gateway and some of medical MEMS based sensors. These MEMS sensors are continuously used for tracking several sorts of physiological indicators. The statistics data handshaking between those medical MEMS sensors and the gateway is found out by using intra- wireless body area network communications, which has been underlined in IEEE stds. 802.15.4 [4] and 802.15.6 [5]. Every gateway gathers biomedical signals from medical sensors which have been distributed or placed in the body, and then onwards those to corresponding medical monitoring centers WBAN communications channels. Although many researches had been carried out on growing WBAN-based health monitoring systems [6]–[7], maximum of them are restrained their attentions on intra-WBAN network communications, even rare studies regarding WBAN communications has been done [8]. Indeed, it is vital to model proficient transmission approach for scheduling beyond-WBAN communications for the following aspects i) the transmissions of data in beyond- wireless body area network would be explosively waned with an extensive distribution of wireless body area network in the next future; and ii) outdated technologies, like cellular networks and WiFi, can barely encounter the needs for pervasive electronic-health services because of their limitations in spectrum resource [9].

In addition, for transmissions of medical data, the most important things to be considered is the quality of service (QoS) being provided for the data transmission. This need a promising indicator that have to report a priority of higher order over non-emergent ones [10]; and ii) transfer of consistent medical data consisting of physical parameters should be planned by conjointly taking the information and delay time [11]. In electronic-health systems, based on the application-specific criticality levels the packets of data are generally categorized into numerous classes. For instance, the data acquired from the electrocardiogram is given priority more significant than body temperature, though both the parameters belonging nonemergency. Nevertheless, excessive delays occur in any generous of information related to medical data may also lower the timeliness of data transmission or distress the accuracy of disease monitoring and diagnosis. Along with it, the implementation of priority-aware beyond wireless body area networks transmission scheduling depend on the medical packets, which is needed to be reported by concomitant gateways. Nevertheless, unlike simple-architecture sensors, wireless body area networks-gateways are meant to be smart devices high intellectual data processing capability, So that they may behave tactically and inconsiderately [11]–[12]. Perceptibly, in fine structured electronic-health architecture, such deceptive behaviors should be banned. If not, the medical-grade quality of services can certainly not be convinced because of the imprecisely stated priority information, which leads to stern concerns in healthcare system. Furthermore, in order to expedite the application of the system, the complete performance of transmissions data of all medical packets should be maximized.

Technically mentioning, the designing this sort of beyond- wireless body area network mechanism for transmission scheduling with all above-mentioned capabilities is quite challenging due to attain medical-grade quality of services, beyond-wireless body area network data transmissions is to be modeled consistent with a delay-based precedence order, in the meantime defensive emergent information deliveries. It necessitates the attention of priorities in dynamic scheme in the data scheduling method. a) Since the gateways for data transformation are coherent and selfish, the modeled scheduling mechanism need to be capable of induce gateways to candidly state the real classes in their medical data packets. b) With dynamic natures of beyond- wireless body area network data transmissions and considered from communication gateways, it is tough to effectively control the scheduling unit in higher order. In the preceding works [13], [14], design approach was adopted firstly in beyond- wireless body area network scheduling the transmission. Nevertheless, there exist some sturdy boundaries in mechanisms contrived by [13] and [14]: i) both the systems laboring a class-based prioritized scheduling g, which may tremendously leads a huge delays in data transmission for certain medical data packets; ii) the time of transmission of various data packets in beyond-WBANs was considered to be both homogeneous in traits or exponentially placed, which are indeed not practical; and iii) none of them aimed to wane the network health, hence the performance of overall beyond- Wireless body area network transmission scheduling can be far from optimal.

To address the aforesaid difficulties and deal with the previous works' limitations, on this survey paper, we mentioned a scheduling mechanism for delay-based priority WBAN transmissions in data. In the taken into consideration version, sensed data of medical triggers produced by body reach randomly at each gateway through Intra-WBAN communications. Upon the reception of a data packet, the gateway without delay reports a beyond- Wireless body area network transmission request to the bottom station (BS)which is then coupled towards medical units via Internet, and momentarily keeps the data packet before it is been scheduled to be transferred. Medical data packets may experience diverse transmission time in beyond- Wireless body area network because of their uncommon nature in the size of data packet and doable SNRs at data transfer gateways. At the controller of network the base station copes the beyond wireless body area network transmissions scheduling with the fulfillment of quality of service requirements at the prescribed standards. For getting the most out of the network welfare whilst guaranteeing that no gateway interface has the inducement to overstate the classes of their data transmission of in the form of medical packets, an efficient and truthful mechanism is deliberated.

1.1 Risks in WBAN Routing Protocol Design:

WBAN is intentionally developed to applied in to or on human, hence this network around is very multifarious. Since the physical attributes of biological data composed have significantly affect human health, it is quite imperative to model transmission protocol for BAN. The below mentioned challenges and problems must be taken into consideration in the design of routing based on the above analysis, [16, 17].

1. Topological Structure and dynamic changes: The transmission of wireless data includes body parameter transmission, and transmission of free space. Effect of shadow occurred in the motion of human would also be considered. With the drive of limbs and other body movement, the comparative point between nodes would also change. Since the time-variation of the topology [18], consistent mechanism for routing would be modeled to acclimatize the dynamic topology.

2. Efficiency: WBAN is a human body attached technology; some memes based bio sensors are put into the human body via surgery. Not enough to power is supplied via micro-batteries. Though Radio Frequency, and Electro Magnetic technology can be utilized for the supplement of power and an efficient design should be supported to maintain the energy management. Therefore, two things are considered such as energy efficiency of single node and energy balance of whole network in the designing of routing topology, and thus extend the life time of those networks.

3. Temperature on Node: Heat is generated at each nodes when these nodes are active, which damage the tissues and organs where it is next to kept in the human body [19, 20]. Hence, the problem of temperature must be encountered in the design of routing, in order to eliminate this situation.

4. Different Quality of Services necessities: Nodes in the wireless body area network cause different varieties of information must differently be processed, to guarantee the quality of services data of kind emergency data, delay sensitive data, reliability sensitive data, and general data are required.

As WBAN is an exceptional type of wireless sensor network, it receives many of its difficulties. Though, a number of challenges characterize WBAN and a number of problems necessitate improved solutions. Adoption of wireless body area network wouldn't be accomplished without engaging in several technical and social challenges such networks faces. The essentiality of user such as security and compatibility, , ease of use, privacy, safety are also of great prominence. This survey contemplates an extensive range of challenges in wireless body are network.

A) Consumption of Power: Gadgets in wireless body area network are by and large battery fueled. WBAN with less and littler hubs with littler batteries contrasted with other wireless sensor network, which includes more limitations power utilization in correspondence. The force required by hubs in wireless body area network fluctuates as indicated by the application type. Every single embedded hub is required to work for numerous years. [2014, Cavallari et al.]ICD, for instance, need to work for in any event 5 years. In this way, it is fundamental to plan ultra-low force radio handsets. Wireless Body Area Network conventions must have the option to limit power utilization without relinquishing unwavering quality. A typical system is to permit gadgets to rest for more often than not and in this way lower the obligation cycle. Be that as it may, adjusting between power utilization and normal start to finish deferral ought to be considered. The principal point to consider while picking a remote innovation for Wireless Body Area Network is the force utilization. Wireless Body Area Network top force requests out of gear mode change somewhere in the range of 0.001mW and 0.1mW and needs up to 30mW dormant mode. Remote innovations center on limiting the normal current drawn from the battery by various methods and systems. In addition upgrades are important to diminish the strained pinnacle current in detecting advances, radio equipment and incorporated circuits. Issue in limiting obstruction and expanding Wireless Body Area Network lifetime by receiving transmit power control necessitate further consideration. A few examinations explored in the writing center around searching vitality from body warmth or movement. Force utilization challenge is profoundly talked about in Movassaghi et al. and Cavallari et al 2014.

B) Devices having Heterogeneity: Sensors in wireless body area network catch various types of information, dependability is a key subject. For example, sensors shift in their detected rush hour gridlock rate relies upon the application type and information which is sent. Rate of bit esteems fluctuate between 1kbps and 10Mbps [2014Cavallari et al.]. Inalienably, a few sensors sense more basic information than others. Besides, a similar sensor may be in various states that fluctuate in their need. Thus, the unwavering quality evaluation may change powerfully at runtime. For instance, temperature of human may be ordinary and needs a typical degree of unwavering quality, yet when the body temperature out of nowhere get high or fall down as far as possible, the dependability prerequisite turns out to be considerably more thorough. As an outcome, WBAN needs to progressively ensure unwavering quality for the sensor hubs. Guaranteeing a unique degree of unwavering quality for various sensors is a test vital. These elements, just as sensor constraints, breakdowns and impedance change the system operational situations, which thusly prompts deficient and incorrect sensor information [Ullah et al. 2012].

C) Reliability: A wireless body area network requests a high level of dependability as it legitimately influences the nature of patient checking. Unnoticed hazardous circumstances can prompt casualty. Along these lines, unwavering quality is a critical issue in WBAN. Dependability can be estimated by the nature of the connection or by the effectiveness of start to finish correspondence. So as to meet client desires and accomplish a dependable system, there are three fundamental qualities that any system innovation, modified help and convention needs to address: adaptation to internal failure, Quality of service and security [2013, C.N.A.P Staff]. Planning conventions for unanticipated issues is a basic component of WBAN structure since it is fundamental that the WBAN works persistently for clients who depend on it. This necessitates the engineering of a WBAN and its plan strategies ought to be deficiency tolerant. A flaw tolerant system constrains the impact of a failure, with the goal that the least number of system segments are influenced. In any case, WBAN clinical applications make better standards for the nature of the conveyed benefits with respect to such applications any consistent breaks, stops, deferrals or parcel misfortune could be deadly particularly in crisis circumstances]. Consequently, Quality of service turns into an ever-expanding necessity of WBAN. Solid QoS is procured by proposing very much planned conventions that can organize arrange traffic. Dependable QoS likewise implies that that all parcels show up on schedule and in their right request. This requires components that keep away from or could oversee traffic clog. System transfer speed quantifies the limit of the medium to convey information that is the measure of transferred data via the channel in specific time.

D) The Threats: Failure, Faults and Error, a risk is an infringement of at least one of the framework prerequisites. A few dangers could be seen without influencing the usefulness of the framework. Others could be seen yet contrarily influence the framework usefulness as they disregard the requirements of the framework. To comprehend the idea of adaptation to internal failure, three kinds of dangers ought to be separated, that is failure, error and fault. Any risk begins as fault; this is a physical equipment imperfection or a product deformity. It could happen deliberately or coincidentally. On the off chance that error is an active fault. In such case the system is not working as it is intentionally made to bel. In the event that the flaw can only with significant effort be seen, at that point it is known as an inactive deficiency, for example, a bug existing in the code. In the event that a flaw happens and stays unconsidered, it is conceivable that it broadens and influences other system parts and subsequently turns into a blunder. The mistake is a seen risk, which, when it happens drives the framework into the condition of carrying on badly [2014Alrajei and Fu]. As it were, a blunder is a functioning shortcoming. In the event that blunders proliferate, they can cause network failure. Right now, network doesn't accomplish the right assistance it should offer. Service is defined as a number of comprehensible system's external states. In such case of the total system goes amiss from the right assistance express. For instance, when a node drains its vitality because of a dead battery deficiency, lost association with different hubs blunder happens. The non-active node is currently not ready to work in the system and different nodes are not ready to get information from it. This implies an error is proliferated inside the system, prompting a variation from the norm in the system conduct, which causes network disappointment. In any case, not all mistakes lead to a general network disappointment as they are not ready to influence the outer condition of the framework, prompting different types of failure that cause different methods of management failure.

2. Literature survey

As in WBAN (wireless body area network) sensors are positioned around or on or into the body, with the small battery. The life of battery is proportional to the size of batter. In WBAN data is analyzed in a systematic power effectual manner. Whilst connive wireless body area network power efficiency is the key feature besides the consumption of high layered energy-efficient media access control protocols. A kind of this protocol is projected by [21] in which multi-hop architecture had deliberated. As the data handled by WBAN, is of less powered, the projected a strategy created on TDMA based MAC protocol for data communication among the sensor unit and gateway. It shows that consumption of power is little in contrast to time division multiple access based Star network.

Another vitality effective model was proposed by [22] which are known as a remote gadget driver for low obligation peripherals. This gadget controls all the gadgets associated with it to make the framework dependable. The gadget driver isn't constantly associated with the peripherals and client can withdraw/rejoin it as per the prerequisite. These procedures spare the dormancy time by utilizing either Bluetooth or Zigbee.[23]

Impact, overhead, bundle overhead, and so forth are the primary purposes behind vitality wastage in the structure of MAC convention in WBAN [24-25].In these two conventions proposed were CSMA/CA and TDMA. For research and usage purposes, they utilized the IEEE standard 802.15.4 (Bluetooth and Zigbee) as correspondence conventions. The Zigbee has a low information rate and low data transmission when contrasted with Bluetooth. Zigbee is utilized with star topology though Bluetooth is broadly utilized with ace slave arrangement.

In [Emi05], the creators of this work presents a multi-level telemedicine framework and depict how they improved our model WBAN usage for PC helped physical restoration applications and wandering observing. The framework performs constant examination of sensors' information, gives direction and input to the client, and can create alerts dependent on the client's state, level of action, and ecological conditions. Furthermore, all recorded data can be moved to clinical servers by means of the Internet and consistently coordinated into the client's electronic clinical record and research databases.

Jocelyne Elias et al. [27] (2012) projected a solid topology structure and provisioning approach for WBAN named as RTDP-WBAN that considers the versatility of the patient while ensuring a dependable information conveyance required to help human services applications' needs. To do as such, they initially proposed 3D organize framework ready to compute the directions of hand-off sensor hubs in various body stances and developments. This framework utilizes a 3D-model of human and a particular arrangement of hub positions with steady correspondence joins, shaping a virtual spine. Then, they explored the ideal transfer hubs situating mutually with solid and financially savvy information directing for various body stances and developments. Along these lines, they utilize an Integer Linear Programming (ILP) model that can locate the ideal number and areas of hand-off hubs and compute the ideal information steering from sensors and transfers to the sink, limiting both the system arrangement cost and the vitality utilization. They settled the model in unique WBAN situations such as Sit, Stand, and Walk, and contrast its exhibition with other transferring methods.

N. Javaid et al. [28] (2013) exhibited a scientific conversation about the vitality proficiency of Medium Access Control (MAC) conventions for WBASN. For this reason, diverse vitality productive MAC conventions with their separate vitality improvement procedures; Scheduled Contention and Time Division Multiple Access (TDMA), and Low Power Listening (LPL) are explained. They additionally systematically looked at way misfortune models for around the body interchanges in wireless body area sensor networks. These three ways of loss assessment situations are recreated in MATLAB and results indicated that way misfortune is more In-body correspondence due to less vitality levels to deal with organs and tissues situated inside the human body. Furthermore, the force model for WBASNs of CSMA/CA and reference point mode is likewise introduced. The outcomes indicated that the intensity of CSMA/CA mode is less when contrasted with signal mode. At long last, they proposed that cross breed mode is progressively valuable to accomplish streamlining in power utilization, which thus brings about high vitality proficiency.

N. Javaid et al. [28] (2013) exhibited a diagnostic conversation about the vitality productivity of Medium Access Control (MAC) conventions for WBASN. For this reason, diverse vitality proficient MAC conventions with their separate vitality advancement strategies; Scheduled Contention, Low

Power Listening, and TDMA, are explained. They additionally logically thought about way misfortune models for On-body; Off-body and In-body interchanges in wireless body area sensor networks. These three ways misfortune situations are reproduced in MATLAB and results demonstrated that way misfortune is more In-body correspondence in view of less vitality levels to deal with tissues and organs situated within the human body. Besides, the force model for WBAN of CSMA/CA and reference point mode is likewise displayed. The outcomes of CSMA/CA mode demonstrated that the intensity is less when contrasted with reference point mode. At long last, they proposed that mixture mode is increasingly valuable to accomplish enhancement in power utilization, which therefore brings about high vitality effectiveness.

A significant necessity in planning transmissions of multiclass clinical parcels with various criticality is the need mindfulness [30].Structured an effective organized medium access convention dependent on IEEE 802.11e for the clinical remote neighborhood. Rezvani et al. in [31] built up a setting mindful versatile asset assignment calculation for WBANs, in which information deals were organized by clinical circumstances and channel conditions. Phunchongharn et al. in [32] presented inventive remote access conspire for medicinal services systems which could maintain a strategic distance from electromagnetic obstruction and give two classes of needs. Other than these, defer limitations for clinical parcel transmissions, as other key element of internet of things-based medicinal services, have been talked about in [33], [34]. In particular, in [33] proposed a vitality proficient rate of transmission adaption approach for deferring obliged continuous medicinal services observing, and in [34] broke down the conjunction of WBAN based on Zigbee and WiFi with the assurance of clinical evaluation defer prerequisites. In any case, none of these works researched effects of gadget knowledge on clinical bundle transmission booking.

System plan strategy has been broadly concentrated in different remote applications for keeping untruthful practices from wise and childish clients [35]–[37]. For example, Xu et al. in [35] displayed a plot safe twofold sale system to decide the ideal hand-off task for expanding the general system throughput. Yi et al. in [36] figured a two-arrange range imparting system for intellectual radio systems to unsure essential range utilizations. Since queueing displaying is a viable device in depicting the dynamic administration of remote transmissions [38], component configuration incorporating queueing booking has been as of late perceived in the writing [39], [40] as a characteristic and incredible approach to investigate the exhibition of dynamic frameworks with private data from people. Nonetheless, every one of these components can't be legitimately applied in light of the fact that the models considered didn't suit the IoT based medicinal services frameworks, and every one of them conventionally disregarded clinical evaluation QoS prerequisites on transmission planning.

In this work [41], we study the issue of how to plan a clinical evaluation remote neighborhood (WLAN) for social insurance offices. To start with, not at all like the IEEE 802.11e MAC, which orders traffic fundamentally by their postpone imperatives, we organize clinical applications as per their clinical earnestness. Second, we propose a component that can ensure total need to each traffic classification, which is basic for clinical evaluation nature of administration (QoS), while the ordinary 802.11e MAC just gives relative need to each traffic classification. At long last, for legitimate execution assessment from a clinical perspective, we present the weighted symptomatic contortion (WDD) as a clinical QoS metric to adequately gauge the clinical diagnosability by separating the principle analytic highlights of clinical sign. Our reenactment result shows that the proposed component, together with clinical classification utilizing total need, can fundamentally improve the clinical evaluation QoS execution over the ordinary IEEE 802.11e MAC.

The improvement of tele-monitoring by means of remote body zone systems (WBANs) is a developing course in customized medication and locally situated portable health [42]. A WBAN comprises of little, smart clinical sensors that gather physiological parameters, for example, EKG (electrocardiogram), EEG (electroencephalography) and pulse. The recorded physiological signs are sent to an organizer by means of remote advancements and are then transmitted to a human services checking focus. One of the most generally utilized remote innovations in WBANs is Zigbee in light of the fact that it is literally focused on applications which generally need a minimum data rate and long standing battery life. In any case, Zigbee-level wireless body area networks face extreme obstruction issues within the sight of WiFi systems. To tackle this issue, we have built up a calculation that have the control in WiFi systems to ensure the postpone necessity for physical signs, particularly for crisis messages, in situations with the conjunction of Zigbee-level wireless body area network and WiFi. Applications of WiFi create traffic with various defer necessities, we center just on WiFi traffic that doesn't have stringent planning prerequisites. Right now, we propose a versatile burden control calculation for Zigbee-level WBAN/WiFi conjunction situations, with the point of ensuring that is experienced by Zigbee sensors doesn't surpass a maximally middle of the road timeframe. Recreation results show that our proposed calculation ensures the postpone execution of WBANs based on Zigbee by relieving the impacts of WiFi obstruction in different situations. Then, the modeling of an efficient and truthful and scheduling strategy for delay-dependent priority-aware medical data transmissions of packet is framed.

Mix of smaller than expected sensors makes a remote body territory organize (WBAN) [43], which empowers remote wellbeing checking. In such cases, visualized clinical and non-clinical uses of WBANs literally have a similar need except if in crisis circumstances. Additionally, clinical utilizations of WBANs want some severe necessities that are not so significant for non-clinical applications, for example, low-power utilization or unwavering quality. Also, divert condition might change in WBANs on account of blurring impacts and this causes bundle misfortune. Consequently legitimate traffic prioritization, high dependability and productive channel use are fundamentally significant issues in these systems. Right now, creators improve the presentation of the MAC convention of WBANs utilizing a versatile asset allotment and traffic prioritization as per the clinical circumstance of client and channel condition. Through isolating adaptively and dealing with the potential deals of wireless body area network, the heterogeneous prerequisites of various applications of the equivalent are given. Expository and reenactment results illustrate the proposed MAC convention beats IEEE 802.15.6 and IEEE 802.15.4 MAC conventions as far as force utilization just as the channel use and unwavering quality.

Current development in wearable device benefits by authorizing prevalent observing of human [44]. Mill wellbeing perceiving framework encompasses of a system of wearable sensors that recurrently monitor physiological parameters. Information collected is handed-off using existing remote correspondence pacts to a base station for additional handling. It provides scientists with data to think about the existing low-power correspondence advances that can conceivably bolster the fast improvement and arrangement of WBAN frameworks, and fundamentally pointed around remote monitoring of older or incessantly unfit patients in private conditions.

Communication adequately upgrades the channel limit of remote systems by permitting some single-radio wire hubs to transfer information for different hubs [45]. In such a correspondence plot, picking suitable hand-off hubs is basic to amplify the general system execution. Right now, consider the task issue of hand-off hubs in an agreeable remote system, where physical hand-off frameworks and transfer supporting services (hand-off task) are autonomously worked by various narrow pathed

IOP Conf. Series: Materials Science and Engineering **983** (2020) 012017 doi:10.1088/1757-899X/983/1/012017

elements, every one of it is determined by its own advantage. We initially figure the issue as a rehashed twofold sale by considering the advantages of all substances in the framework. It consider a framework comprising of a lot of source-to-goal sets, transfer hubs, bunch specialists, and the barker, where source hubs are assembled into various gatherings and each gathering is spoken to by a gathering operator. The source hubs and gathering operators look for chances to amplify their own advantages through untruthful offering, plotting with one another, etc. We at that point show that these practices will endanger the social advantage of all substances in the framework. Experimental outcomes show that the proposed sell off is compelling in conspiracy opposition with limited intrigue probabilities. To our best information, this is the first sale instrument for transfer task in quite a while that is honest, conspiracy safe, spending plan parity and individual-reasonable.

An on-request service stage interfaces holding up time-delicate clients with free specialist co-ops operators [46]. This paper inspects how two characterizing highlights of an on-request administration stage defer affectability and operator freedom sway the stage's ideal per-administration cost and compensation. Defer affectability diminishes anticipated utility for clients and operators, which proposes that the stage ought to react by diminishing the cost to support the investment of clients and expanding the compensation to empower the interest of specialists. This natural cost and pay remedies are legitimate in a benchmark-setting without vulnerability in the clients' valuation or the operators' chance expenses. Be that as it may, vulnerability in either measurement can turn around the remedies: Delay affectability builds the ideal cost when client valuation vulnerability is moderate. Under specialist opportunity cost vulnerability, operator freedom diminishes the cost. Under client valuation vulnerability, operator freedom builds the cost if and just if valuation vulnerability is adequately high.

3. Delay-constrained priority queueing model

In this section, it studies about the queueing models of data transmission and network of the beyond-wireless body area network under consideration. Consider a network beyond-wireless body area network [17], [18] with a single base station and M gateways, each has been used for a patient, as shown in Figure 1. There are N similar channels of radio that are devoted for beyond-wireless body are network data transmittance which is generally speaking transmissions of data from gateways to the base station, and the base station is accountable to schedule uplink for data transmissions from all other communication gateways. Respective communication gateway may obtain different medical data packets engendered by its coupled medical bio sensors node via intra-wireless body area network data transmittance. Conferring to IEEE 802.15.6 [6], data packets are generalized as a finite set of data classes that contains classes of emergency and non-emergency for emergency single class data is furthered and multiple classes for non-emergency.

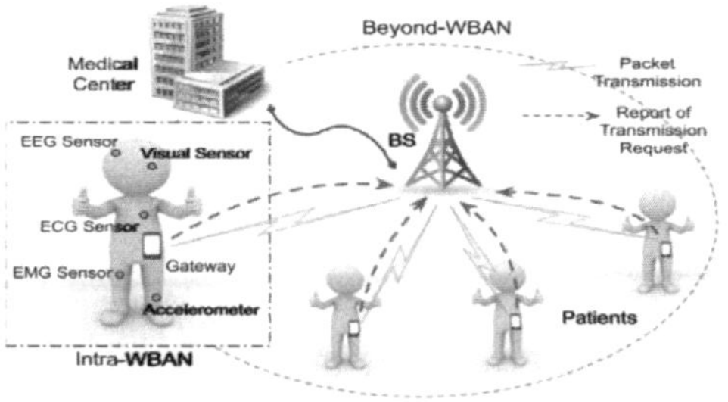

Figure 1.Electronic-health networks

Instinctively, dissimilar medical data packets is heterogeneous in terms of their size of packet, by assumption the size in terms of bits of medical data packets in respective class is considered by a arbitrarily random variable succeeding with overall PDF distribution and a finite mean.

More than hundreds of heterogeneous medical sensors are assimilated into the each WBAN, the collective gathering of medical data packets poised from sever a standalone sensors at each communication gateway, could be well estimated as a process of poison with rate. Moreover, via a long standing monitoring, it is supposed that there a known distribution of medical data packets from various classes at respective communication gateway, where the constituent parameters, specifies the likelihood which an attained medical data packet at communication gateway which have its place to class. Thus, the base station (BS) can extravagance all requests of data transfer from a particular communication channel, as stated in Fig. 2, which contains two explicit virtual buffers, namely one for non-emergent ones and the other one for emergent packets.

Note that despite the fact that every single clinical bundle are touchy to the potential holding up delays in past wireless body area network transmissions, there is a crucial prerequisite in human services applications that the basic data conveyance must be ensured with a carefully higher need over those with standard significance. This suggests a natural limitation that the Quality of service of beyond wireless body area network transmissions for new data packets needs to consistently be ensured independent of the presence of non-emergent ones.

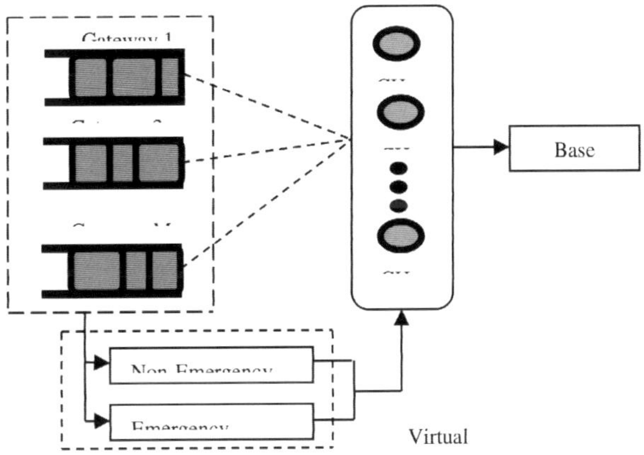

Figure 2.Transmission Scheduling and Priority Queueing model

There are still the existence of delays in waiting in beyond-wireless body area network transmissions, because of the restricted radio channels for beyond-WBANs, Since transmissions of medical data packet information are very much delay sensitive, i.e., with the rise of its waiting delay the value of data transmitting of a medical packet decreases. A parameter θ is defined such that the cost per unit for data packet waiting delay in medical packets data in every classes, which is delay sensitivity of the ith class packets. It is instinctive, for medical data packet, the class of data it goes to is an information of private in nature and its associated gateway is known about it, but undefined to the BS. On the reception of a medical packet, the related communication gateway will assert are quest of beyond-Wireless body area network transmission to the BS with the corresponding packet class. A smart gateway might intentionally misrepresent another class if and only if it can advantage from such behavior as a rational entity, which shows diminishing its cost in the beyond-WBAN transmission.

3.1 Truthful Mechanism for Transmission Scheduling:

A straight method for arranging medical data packet information transmissions with TMDC delay constraints in internet of things on healthcare networks is modeled. The throughput of the priority based delay constrained data queueing scheme is studied in the first. The individualities of the mechanism system are examined. Based on this mechanism, an efficient rule based pricing is planned and the preferred possessions of TMDC are examined.

The studying of beyond-wireless body area network data transmission scheduling method with data queueing disciplines still very problematic due to i) each packet of medical data has got delay limit hence the potential data packet missing due to displeased services which has to be considered with the service; and ii) there are several data transmission channels are a accessible for beyond-wireless body area network, which demands the analysis of a multi-tier queueing mechanism. In order to scrutinize this complex queueing mechanism system so as to gain the probabilities of data packet loss, model a Markov chain to define processes of serving and the dropping data queue mechanism; and it is subjected to compute the probabilities of data packet transmission/loss by recursions. The flow of time division multiple accesses is listed as follows.

1) Internet of things-oriented healthcare model network is given; the BS articulates and resolves a system model for delay-constrained beyond-wireless body area network data scheduling on packet transmission.
2) Base station chooses to service a non-preemptive priority queueing system and an effectual rule of cost, for waxing the probable revenue R, while assuring quality of services in medical-grade, truthfulness and specific rationality.
3) The BS transmissions the resolute time division multiple access to all related communication gateways.
4) Once a communication gateway gathers a medical data from its biomedical biosensors, this will instantly send a beyond-wireless body area network (WBAN) request for transmission to the base station by mentioning the level of priority for this packet.
5) The base station accomplishes the longstanding beyond-wireless body area network TMDC data transmission scheduling.
6) All packets of data are put in storage in communication gateways' buffers previously scheduling for transmission of data.

4. Applications

Wireless body area network have great latent for many related applications which includes internet of things related medical analysis, cooperative gaming, and military communication.Table.1demonstratefewin and on body applications of WBAN. Observing and program changes for pacemakers and ICD functions bladder controller, and reinstatement of the movement of limbare in-body applications. On-body applications embrace analyzing heart rate, temperature, blood pressure, and respiration. Non-medical on-body application comprise observing things not yet considered, forming of social network, and assessment of soldier tiredness and scuffle readiness. Some applications of the WBAN are itemized in the following part of discuss.

A. Cardiovascular Diseases: A WBAN is a prominent technology to avert the incidence of heart break, analyzing episodic events of unsteady triggers or other irregular condition is utilized for monitoring health

B. Detection of Cancer: The major intimidation to the life of human. Conferring to Health Statistics, around 9 million are affected and is detected during diagnosis of cancer in 1999 [43] and this surges each year. A set of miniaturized bio-sensors capable of observing cancer cells would be impeccably assimilated in wireless body area network. This allows doctors to analyze tumors with no biopsy.

C. Telemedicine Systems: Current telemedicine systems utilize special wireless data channels to exchange data and information to the distant monitoring and analyzing stations, or protocols of low power such as Bluetooth, which are vulnerably open to intrusion by other category of devices employed in working in the similar band of frequency. These limit characteristics protracted health diagnosing. A wireless body area network can be combined into a tele-medicine system which cares inconspicuous

Table 1. Shows demonstrate some of the WBAN applications

Application Type	Sensor Nodule	Data Rate	Duty Cycle (per device)% per time	Consumption of Power	Sensitive in terms of Latency QOS	Status of Privacy
	Sensor for Glucose	Some kbps	<1.00%	Extremely Low	True	Maximum
	ICD	Some kbps	<1.00%	Minimum	True	Maximum
	Capsule for Endoscope	>2.00 Mbps	<50.00%	Minimum	True	Medium
	Electrocardiogram	3.00 Kbps	<10.00%	Minimum	True	Maximum
	SpO2	32.00 bps	<01.00%	Minimum	True	Maximum
	BP	<10.01 bps	<01.00%	Maximum	True	Maximum
On the body Non-Medical	Headsets for Music	1.400 Mbps	Maximum	Relatively Maximum	True	Minimum

health analyzing for some duration of time.

E. Battlefield: Wireless body area network can be used to into the body of soldiers in a combat zone and monitor their doings to the respective commander, such as firing, running and digging. Though, the soldiers are protected communication pathway in order to avert ambushes.

5. Conclusion

This survey, the organization of priority based on delay-dependent aware beyond-wireless body area network medical data packet data transmissions in electronic-health system has been well considered in all perspective it matters. In order to describe the active nature in beyond-wireless body area network data scheduling system for data transmission, a multi-tier queueing mechanism with a normally dispersed delay-dependent and time based priority discipline is expressed. The anticipated system not only promise all communication gateways to list the requests of data transfer of their bio medical data packets with actual data, but also diminish the probable costs of delay of the complete system associated to other associated systems. The scope of the work is to implement extra common data packet advent processes and scrutinize the beyond-wireless body area network transmission scheduling without actual expressions of waiting delays. As well, the efficiency of medical information exist for a restricted intervals of time, it is essential to study how that delay-constrained necessities is assimilated in the data queuing mechanism.

References

[1] Ullah S et al 2012 A comprehensive survey of WBANs *J. Med. Syst* **36** 3 1065–1094.

[2] Cavallari R et al 2014 A surveys on WBANs: Technologies and design challenges *IEEE Commun. Surveys Tuts* **16** 3 1635–1657.

[3] Zuhra F T et al 2017 Routing protocols in WBANs: A comprehensive survey *J. Netw. Comput. Appl* **99** 73–97.

[4] IEEE Standard for Local and Metropolitan Area Networks—Part 15.4 *(LR-WPANS) Amendment 4: Alternative Physical Layer Extension to Support (MBAN) Services Operating in the 2360 MHz 2400 MHz band* 2013 IEEE Standard 802.15.4j-2013 pp 1–24

[5] IEEE Standard for Local and MAN—Part 15.6: *WBANs*, 2012, IEEE Standard 802.15.6-2012, pp 1–271

[6] Moosavi H et al 2016 Optimal relay selection and power control with quality-of-service provisioning in WBANs *IEEE Trans. Wireless Commun* **15** 8 5497–5510.

[7] Manirabona A Fourati L C and Boudjit S 2016 A priority-weighted round robin scheduling strategy for a WBAN based healthcare monitoring system *in Proc 13th IEEE Annu. Comm. Net. Conf* 224–229.

[8] Movassaghi S Smith D Jamalipour A Lipman J and Abolhasan M 2014 WBANs: A survey *IEEE Comm. Sur. Tuts* **16** 3 1658–1686.

[9] Niyato D et al 2009 Remote patient monitoring service using heterogeneous wireless access networks: Architecture and optimization *IEEE J. Sele. Areas Comm* **27** 4 412–423.

[10] Lee H Choi C H Ko Y B and Park K J 2011 Wireless LAN with medical grade QoS for e-healthcare *J. Comm. Netwo* **13** 2 149–159.

[11] Misra S et al 2015 Priority-based time-slot allocation in WBANs during medical emergency situations: An evolutionary game-theoretic perspective *IEEE J. Biomed. Heal. Info* **19** 2 541–548.

[12]Jiang T Wu S and Wang H 2017 Interference mitigation for wireless body area networks with fast convergent game *Proc. IEEE Glob. Telecomm. Conf* 1–6.

[13] Yi C et al 2016 An incentive-compatible mechanism for transmission scheduling of delay-sensitive medical packets in e-

health networks *IEEE Trans. Mob. Comp* **15** 10 2424–2436.

[14] Yi C and Cai J 2017 A priority-aware truthful mechanism for supporting multi-class delay-sensitive medical packet transmissions in e-health networks *IEEE Trans. Mobile Comput* **16** 9 2422–2435.

[15] Zhang K Yang K Liang X Su Z Shen X and Luo H H 2015 Security and privacy for mobile healthcare networks: From a quality of protection perspective *IEEE Wirel.Comm* **22** 4 104–112.

[16] Yessad N Bouabdallah A Tari A and Omar M 2017 *QoS-based routing in WBANs: A survey and taxonomy. Computing* 100: 245–275 doi: 10.1007/s00607-017-0575-4 [CrossRef] [Google Scholar]

[17] Amit S and Sudip M 2018 Dynamic Connectivity Establishment and Cooperative Scheduling for Qos-Aware WBANs *IEEE Trans. Mob. Compu* **17** 2775–2788.

[18] Yuan X Shen X S Ye Q Zhang K Li C Cheng N and Zhang N 2018 *Performance Analysis of IEEE 802.15.6-Based Coexisting Mobile WBANs with Prioritized Traffic and Dynamic Interference*. [IEEE Trans. Wirel. Comm, vol 17] pp 5637–5652 doi: 10.1109/TWC.2018.2848223

[19] Roy M Chowdhury C and Aslam N 2017 *Designing an energy efficient WBAN routing protocol* [Proceed. of the Internet (COMSNETS) Bangalore] India 4–8 pp 298–305 [Google Scholar]

[20] Jiang W Miao T T Wang Z C and Feng M 2017 *A Survey of Thermal-Aware Routing Protocols in WBANs* [Proc. of the IEEE Intern. Confe. on CSE and IEEE ntern. Confe. on Embed. and Ubiqu. Comput. (EUC) Guangzhou China 21–24] pp 17–21 [Google Scholar]

[21] Nabi M 2010 *A Robust Protocol Stack for Multi-hop WBANs with Transmit Power Adaptation* in 5th Ann. [Intern. ICST Confe. on Body Area Netw. Corfu Island Greece]

[22] Yan L 2007 *Energy Comparison and Optimization of WBAN Technologies* [Proceed. of the ICST 2nd Inter. Confer. on BAN (Body Nets '07) Florence Italy] 1-8.

[23] http://www.zigbee.org

[24] Ullah S 2012 *A Study of MAC Protocols for WBANs*,[(IJASUC) **3.**

[25] Wei Ye J Estrin D and Heidemann 2002 An Energy-Efficient MAC protocol for WSNs *INFOCOM 2002. Twenty-First Ann. Joint Confer. of the IEEE Comp. and Comm. Soci.. Proceed. IEEE* **3** 1567 – 1576

[26] Beom Seok Kim Jongbum Ryou Jinsung Cho and Ben Lee 2011 *An Emergency Handling Scheme for Super frame structured MAC protocols in WBAN* [5th Intern. Confere. on Ubiqui. Informa. Manag. and Commun. (ICUIMC2011) February] pp 21–23

[27] Elias J Jarray A Salazar J Karmouch A & Mehaoua A 2012 *A Reliable Design of WBANs situations* vol *3*

[28] Javaid N et al 2013 *Energy Efficient MAC Protocols in WBASNs-A Survey* arXiv preprint arXiv *1303.2072*

[29] Ahmed S Qasim U Akbar M Javaid N Khan A and Iqbal Z A LAEEBA: (AINA) 2014 *[IEEE* 28th Inter.l Conf.] pp 435-440

[30] Rashwand S and Misi ̌ c J 2012 Two-tier WBAN/WLAN healthcare networks: priority considerations *in Proc. IEEE GLOBECOM* 5398–5403.

[31] Rezvani S et al 2013 Context aware and channel-based resource allocation for WBANs *IET Wireless Sens. Sys* **3** 1 16–25.

[32] Phunchongharn P et al 2010 An EMIaware prioritized wireless access scheme for e-health applications in hospital environments *IEEE Trans. Infor. Technolo. Biomedic* **14** 5 1247–1258.

[33] Liu Z Chen C W Liu B and Chen C 2015 Energy-efficient resource allocation with QoS support in WBANs *in Proc. IEEE GLOBECOM* 1–6.

[34] Kim Y et al 2016 Coexistence of ZigBee-based WBAN and WiFi for health telemonitoring systems *IEEE J. Biome. Heal. Inform* **20** 1 222–230.

[35] Xu Z & Liang W 2014 Collusion-resistant repeated double auctions for relay assignment in cooperative networks *IEEE Trans. Wireless Comm* **13** 3 1196–1207.

[36] Yi C and Cai J 2014 Two-stage spectrum sharing with combinatorial auction and stackelberg game in recall-based cognitive radio networks *IEEE Trans. Communi* **62** 11 3740–3752.

[37] Yi C Zhang G & Cai J 2017 Spectrum auction for differential secondary wireless service provisioning with time-dependent valuation information *IEEE Transa. Wireless Communi* **16** 1 206–220.

[38] Chen Y Wang X and Cai L 2016 On achieving fair and throughput optimal scheduling for TCP flows in wireless networks *IEEE Trans. Wireless Communi* **15** 12 7996–8008.

[39] Mitra M 2001 Mechanism design in queueing problems *Econ. Theory* **17** 2 277–305.

[40] Afeche P 2013 Incentive-compatible revenue management in queueing systems: optimal strategic delay *Manufacturing Servi. Opera. Managemen* **15** 3 423–443.

[41] Lee H Park K J Choi C Hand Ko Y B 2011 Wireless LAN with medical-grade QoS for e-healthcare *J. Communi. and Netwo* **13** 2 149–159.

[42] Rashwand S & Misiˇ c J 2012 Two-tier WBAN/WLAN healthcare networks: priority considerations *Proc. IEEE GLOBECOM* 5398–5403.

[43] Rezvani S & Ghorashi S A 2013 Context aware and channel-based resource allocation for WBANs *IET Wireless Sens. Syst* **3** 1 16–25.

[44] Liu Z Liu B Chen C & Chen C W 2015 Energy-efficient resource allocation with QoS support in WBANs Proc. *IEEE GLOBECOM* 1–6.

[45] Xu Z & Liang W 2014 *Collusion-resistant repeated double auctions for relay assignment in cooperative networks IEEE Trans. Wireless Communic* **13** 3 1196–1207.

[46] Afeche P 2013 Incentive-compatible revenue management in queueing systems: optimal strategic delay *Manufacturing Serv. Opera. Manag* **15** 3 423–443.

AUTHOR INDEX